Analysis of Drugs and Metabolites by Gas Chromatography–Mass Spectrometry

ANALYSIS OF DRUGS AND METABOLITES
BY GAS CHROMATOGRAPHY-MASS SPECTROMETRY

VOLUME 1: Respiratory Gases, Volatile Anesthetics, Ethyl Alcohol, and Related Toxicological Materials

VOLUME 2: Hypnotics, Anticonvulsants, and Sedatives

VOLUME 3: Antipsychotic, Antiemetic, and Antidepressant Drugs

VOLUME 4: Central Nervous System Stimulants

VOLUME 5: Analgesics, Local Anesthetics, and Antibiotics

OTHER VOLUMES IN PREPARATION

ANALYSIS OF DRUGS AND METABOLITES BY GAS CHROMATOGRAPHY–MASS SPECTROMETRY

VOLUME 1

Respiratory Gases, Volatile Anesthetics, Ethyl Alcohol, and Related Toxicological Materials

Benjamin J. Gudzinowicz
Department of Pathology
Rhode Island Hospital
Providence, Rhode Island

Michael J. Gudzinowicz
Center in Toxicology
Department of Biochemistry
School of Medicine
Vanderbilt University
Nashville, Tennessee

With the Assistance of

Horace F. Martin
Department of Pathology, Rhode Island Hospital
Providence, Rhode Island
and
Division of Biological and Medical Sciences
Brown University, Providence, Rhode Island

and

James L. Driscoll
Department of Pathology, Rhode Island Hospital
Providence, Rhode Island

MARCEL DEKKER, INC., New York and Basel

Library of Congress Cataloging in Publication Data

Gudzinowicz, Benjamin J
Analysis of drugs and metabolites by gas chromatography--mass spectrometry.

Includes indexes.
CONTENTS: v. 1. Respiratory gases, volatile anesthetics, ethyl alcohol, and related toxicological materials.
1. Drugs--Analysis. 2. Gas chromatography. 3. Mass spectrometry. 4. Drug metabolism. I. Gudzinowicz, Michael J., joint author. II. Title. [DNLM: 1. Drugs--Analysis. 2. Chromatography, Gas. 3. Spectrum analysis, Mass. QV25 G923a]
RS189.G83 615'.1901 76-56481
ISBN 0-8247-6576-1

MARCEL DEKKER, INC.
270 Madison Avenue, New York, New York 10016

Current printing (last digit):
10 9 8 7 6 5 4 3 2 1

PRINTED IN THE UNITED STATES OF AMERICA

Dedicated to

HELEN L. GUDZINOWICZ
a devoted and understanding wife and mother

PREFACE

In the past two decades, remarkable progress has been made in the analysis of drugs, pharmaceuticals, and related toxicological materials. In great measure, these notable advances can be attributed to technological advancements in two specific types or areas of analytical instrumentation; namely, gas chromatography and integrated gas chromatography-mass spectrometry.

Since James and Martin revealed to the scientific community their gas chromatographic technique which permitted the separation of fatty acid mixtures into their individual components, the rapid growth of gas chromatography has been very evident. This remarkable progress can be directly correlated with the improvements that we have witnessed over the years in gas chromatographic stationary phase, carrier gas, column, and temperature- and pressure-controlling technology. Furthermore, it has assumed a position of even greater analytical significance since the advent of highly specific, rapid, sensitive detection systems.

On the other hand, the integrated GC-MS analytical system is rather unique and exceptional in that it combines the mass spectrometer's unexcelled identification potential with the gas chromatograph's separation capabilities. Although the integration of GC and MS was first reported in 1957 by Holmes and Morrell, it nevertheless remained a dormant, costly, and seemingly unappreciated technique until 1970. Since then, with improved instrumentation at a more reasonable price and newly developed operating techniques, numerous publications have appeared in the literature showing its applicability to a wide variety of difficult analytical problems, thus opening up new horizons for analytical research in toxicology, biochemistry, pharmacology, forensics, medicine, etc. To be able to monitor a drug, its persistence and metabolic fate in biological fluids of man via mass fragmentography at picogram concentration levels provides the researcher with a tool of immeasurable significance.

Because much has been written over the years about the analysis of drugs and their metabolites by either or both techniques, the objectives of these volumes are several-fold: (1) to compile from existing literature in a chronological manner the various GC and/or GC-MS procedures available for the analysis of specific drugs and their metabolites, (2) to describe with as much detail as possible all procedures (qualitative and quantitative) in order that they might be reproduced faithfully in one's laboratory, and (3) to indicate, wherever possible, not only the results, precision, accuracy, and limits of detection achieved by a given procedure, but also its applicability to pharmacokinetic studies. For this reason, in addition to the text, which is well referenced in each section, many illustrations of actual applications and tables of data for each instrumental technique are included as aids to the analyst for his greater appreciation and understanding of the limitations as well as potentials ascribed to each method. As stated in the past, from an analytical chemist's point of view, it is hoped that this deliberately combined visual and factual approach will find acceptance by the reader who would otherwise rely only on his interpretation of the written word relative to some published procedure.

Without wishing to be repetitious, in retrospect it must be again stated that this volume really represents the end result of many tedious and arduous investigations by numerous eminent scientists whose research efforts have appeared in the literature throughout the world. I am indeed humbly indebted to them, and to those journals, publishers, and organizations that granted special copyright permission to the authors.

Benjamin J. Gudzinowicz

CONTENTS

Preface		v
Chapter 1.	RESPIRATORY GASES, VOLATILE ANESTHETICS, AND RELATED TOXICOLOGICAL MATERIALS	1
I.	Introduction	1
II.	Respiratory and Blood Gases	3
III.	Volatile, Low-molecular-weight Anesthetics	38
IV.	Sterilizing Agents, Common Organic Solvents in Blood, and Riot-control Aerosol Irritants	63
	References	75
	General References for GC/MS Drug Analysis	80
Chapter 2.	ETHYL ALCOHOL AND VOLATILE TRACE COMPONENTS IN BREATH, BODY FLUIDS, AND BODY TISSUES	85
I.	Introduction	85
II.	Ethyl Alcohol	86
III.	Volatile Constituents in Human Breath, Fluids, and Tissues	153
	References	180
Author Index		191
Subject Index		207

CONTENTS OF OTHER VOLUMES

Volume 2 HYPNOTICS, ANTICONVULSANTS, AND SEDATIVES

Chapter 1. Hypnotics, Anticonvulsants, and Sedatives: Barbiturate Compounds

Chapter 2. Hypnotics, Anticonvulsants, and Sedatives: Nonbarbiturate Compounds

Chapter 3. Hypnotics, Anticonvulsants, and Sedatives: Nonbarbiturate Compounds, Continued

Volume 3 ANTIPSYCHOTIC, ANTIEMETIC, AND ANTIDEPRESSANT DRUGS

Chapter 1. Antipsychotic and Antiemetic Drugs: Phenothiazine, Butyrophenone, and Thioxanthene Derivatives

Chapter 2. Antidepressant Drugs: Monoamine Oxidase Inhibitors, Tricyclic Antidepressants, and Several Related Compounds

Volume 4 CENTRAL NERVOUS SYSTEM STIMULANTS

Chapter 1. Amphetamines, Xanthines, and Related Compounds

Chapter 2. Phenylethylamine-, Tryptamine-, and Propanolol-Related Compounds

Volume 5 ANALGESICS, LOCAL ANESTHETICS, AND ANTIBIOTICS

Chapter 1. Narcotics, Narcotic Antagonists, and Synthetic Opiate-like Drugs

Chapter 2. Antipyretic, Antiinflammatory, and Antihyperuricemic Agents, Local Anesthetics, and Antibiotics

OTHER VOLUMES IN PREPARATION

Analysis of Drugs and Metabolites by Gas Chromatography–Mass Spectrometry

Chapter 1

RESPIRATORY GASES, VOLATILE ANESTHETICS, AND RELATED TOXICOLOGICAL MATERIALS

I. INTRODUCTION

Because gas chromatography's greatest attraction lies in its economy and versatility, this type of instrumentation has been given considerable attention in the past decade as a means of monitoring the concentration of anesthetics in mixtures expired by anesthetized patients during the course of an operation [1] and has been applied to the determination of the absorption, distribution, and elimination of such compounds as diethyl ether [2-4]. Under usual conditions the gases likely to be encountered are oxygen, nitrogen, nitrous oxide, diethyl ether, carbon dioxide, cyclopropane, chloroform, trichloroethylene, and fluothane (halothane), in addition to other chloro-fluoro-organics.

Using columns of dimethyl sulfoxide and dinonylphthalate connected in parallel with a thermistor detector, Adlard and Hill [5] showed in 1960 that ether, fluothane, oxygen, nitrous oxide, carbon dioxide, and cyclopropane could be resolved in 4.5 min.

Similar separations can be accomplished with parallel columns of 20% (w/w) Narcoil on 60-80 mesh Chromosorb P (3 ft x 1/8 in. SS (stainless steel) column maintained at 70°C) and 25% (w/w) dimethyl sulfoxide on 70-80 mesh Anakrom AS (20 ft x 1/4 in. copper column held at room temperature) using either a thermal conductivity (Fig. 1.1) or cross-sectional ionization (Fig. 1.2) detection system [6].

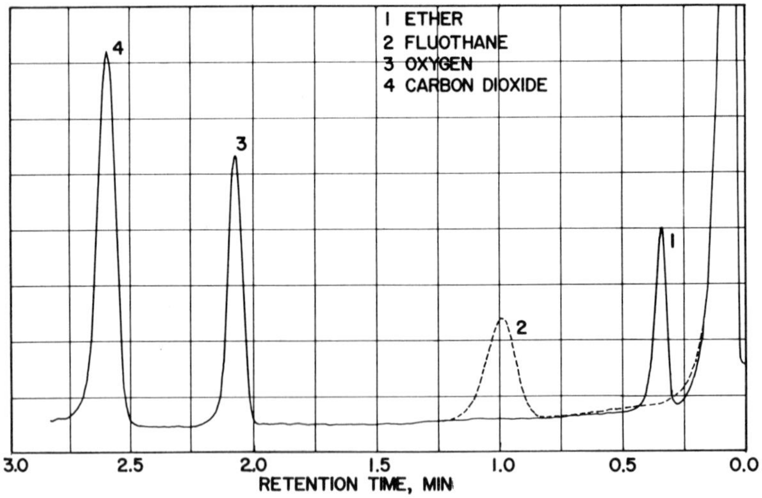

Figure 1.1. Separation of ether, fluothane, oxygen, and carbon dioxide (thermal conductivity cell). From Gudzinowicz [6].

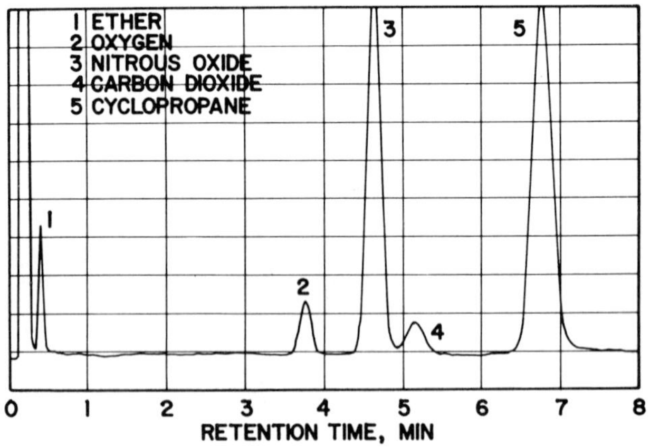

Figure 1.2. Separation of ether, oxygen, nitrous oxide, carbon dioxide, and cyclopropane (cross-section detector). From Gudzinowicz [6].

The basic method of analysis of dissolved gases (O_2, N_2, CH_4, CO, CO_2, etc.) in aqueous systems at concentration levels as low as 0.3 ppm in 0.5-liter samples [7] has been applied to the determination of fluothane tension in blood [8], toxic gases in tissues by direct sampling of air taken in a sealed container [4], and rapid screening of blood samples for volatile substances such as ether, fluothane, chloroform, carbon tetrachloride, ethanol, acetone, trichloroethylene, and paraldehyde [9]. The gas chromatographic approach has also been used for the determination of partition coefficients of volatile anesthetics in blood, tissues, and lipids [10], the apparent volume of distribution, pulmonary clearance, and bioavailability of volatile or gaseous compounds after peripheral intravenous dosing [11], the protein binding of volatile and gaseous compounds such as fluorocarbon aerosol propellants, halothane (fluothane), and cyclopropane [12], sterilizing agents [13,14], common organic solvents in blood [15], and irritant compounds, some of which have been used by military and law enforcement agencies [16-18]. Some of the many methods appearing in the literature will be discussed in this chapter, thus showing the applicability of gas chromatography (hence, integrated gas chromatography-mass spectrometry) to anesthetic and related areas of research.

II. RESPIRATORY AND BLOOD GASES

The approach to respiratory and blood gas analysis is very similar to the determination of the constituents of air important in nitrogen fixation, photosynthesis, as well as respiration--namely, nitrogen, oxygen, carbon dioxide, and water vapor.

In 1958, Smith, Swinehart, and Lesnini [19] showed that gas mixtures containing nitrogen, nitrous oxide, nitric oxide, carbon monoxide, and carbon dioxide could be resolved using a column packed with two layers of silica gel separated by iodine pentoxide. A 10-ft x 1/4-in. length of copper tubing was half-filled with silica gel followed by nearly 8 in. of iodine pentoxide powder and about 1/2 in. of silver metal powder. Additional silica gel was added to fill the remaining length of tubing. At elevated temperatures, the carbon monoxide and nitric oxide were totally oxidized to carbon dioxide [20] and nitrogen dioxide [21], respectively, by iodine pentoxide with the liberation of iodine. Using helium as carrier gas at a flow rate of 30 ml/min, a column temperature of 115°C and a thermal conductivity detector, no separation of carbon monoxide, nitric oxide, and nitrogen occurred in the first section of silica gel, but the carbon dioxide was preferentially retarded. On the other hand, nitrogen was unaffected by iodine pentoxide, but carbon monoxide, when converted to carbon dioxide, lagged behind and was thus separated. The iodine and nitrogen dioxide produced at the iodine pentoxide layer remained in the column; it was postulated that the iodine was apparently removed by the copper tube or the silver metal whereas the nitrogen dioxide was irreversibly adsorbed by the silica gel. The concentrations of nitrogen, carbon monoxide (as carbon dioxide), and carbon dioxide were established from peak height measurements while the nitric oxide was determined by difference. At these specified GC conditions, nitrogen, carbon monoxide, nitrous oxide, and carbon dioxide had retention times of 2.8, 6.7, 9.3, and 10.3 min, respectively, but, as noted by the authors, gas samples containing both nitrous oxide and carbon dioxide could not be adequately analyzed with a column of this length. However, if either one was present alone with the other gases, the method proved satisfactory.

A charcoal adsorption column is inadequate for this type of mixture, because CO_2 is completely adsorbed, and further complications arise due to the reactivity of the packing with nitric oxide, which produces some nitrogen and carbon-oxygen surface complexes [22]. Also, charcoal does not separate nitric oxide and carbon monoxide. Szulczewski and Higuchi [23] described a similar GC separation of these same gases using a dehydrated silica gel column. In their method, after the emergence of carbon monoxide, the column temperature is raised from Dry Ice temperature to room temperature for the elution of nitrous oxide and carbon dioxide. Of greater importance is the fact that nitric oxide emerges as a peak, but several disadvantages are inherent with this method: the long time required for analysis and the rather diffuse separation of the gases which have long retention times. In contrast to these mixtures, the separation of respiratory gases is less complicated because these normally contain inert materials.

For anesthetic research, Hill [1] found it convenient to divide the components to be analyzed into two groups: (1) oxygen, nitrous oxide, and carbon dioxide; and (2) ether, halothane, chloroform, and trichloroethylene. For the oxygen-nitrous oxide-carbon dioxide mixture, this could be separated by a partition column 20 ft long x 1/2 in. i.d. packed with 52-60 B.S. mesh Silocel firebrick impregnated with 20% by weight of dimethyl sulphoxide. In contrast to the majority of gases, including oxygen and nitrogen, which have insignificant solubility in dimethyl sulfoxide (DMSO), certain gases such as SO_2, NH_3, acetylene, CO_2 and N_2 exhibit high DMSO solubilities. With the column maintained at 20°C (room temperature) and using either hydrogen or helium (the latter preferred in the operating theater to reduce explosion risks) as carrier gas at a flow rate of 30 ml/min, a mixture consisting of 3% (v/v) carbon dioxide, 20% (v/v) oxygen, and 77% (v/v) nitrous oxide can be resolved.

To separate the second group of compounds, a 2-ft x 1/2-in. i.d. column packed with 52-60 B.S. mesh Silocel firebrick coated with 15% (w/w) dinonyl phthalate was used with hydrogen as carrier gas (30 ml/min flow rate). At a 20°C column temperature, cyclopropane samples could be injected at 30-sec intervals to obtain its "washout" curve. Operated in this fashion, the washout curve provides a measure of the speed with which a gas or vapor is transported by the blood from the tissues to the lungs and excreted in the breath.

If the anesthetic mixture contains air, oxygen and nitrogen can be separated by gas-solid chromatography with a 12-in. column packed with Linde-type 5A molecular sieve. At a hydrogen flow rate of 60 ml/min, oxygen and nitrogen can be fully resolved in 30 sec. If used in parallel with a 6-ft-long column of 52-60 B. S. mesh Davison grade 70 silica gel (average pore diameter 140 Å), one obtains a chromatogram such as that shown in Figure 1.3 for a mixture of oxygen, carbon dioxide, and cyclopropane. The peaks observed in the chromatogram are deflected in opposire directions since the gas emerging from each of the two columns passes over a thermistor in opposite arms of the Wheatstone bridge.

Figure 1.4 results from monitoring the output of a standard Boyle-type anesthetic machine using an automatic sampling valve. Using a simple ionization chamber detector as suggested by Boer [24], Hill was able to determine nitrogen washout curves with samples injected at 10-sec intervals.

Some practical applications of gas chromatography in the analysis of respiratory gas include a method for measuring carbon monoxide diffusing capacity and lung capillary blood flow [25]. To measure diffusion capacity, the neon, oxygen, and carbon monoxide content of expired air must be analyzed; whereas lung capillary blood flow is based on concentration measurements of neon and acetylene. The analysis of fixed gases at the following concentrations,

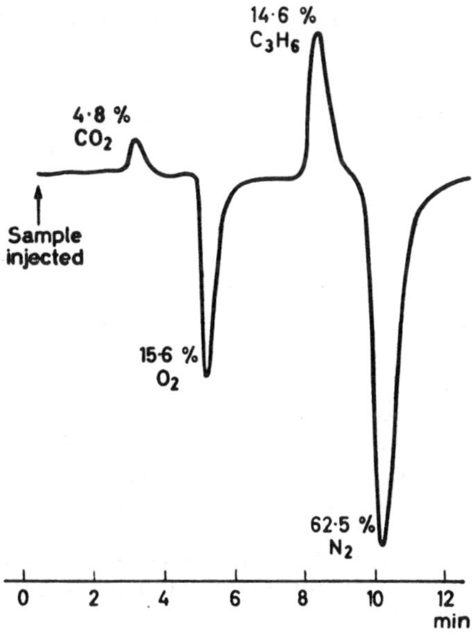

Figure 1.3. Analysis of the mixed expired gas of a dog using a molecular sieve column in parallel with a silica gel column. From Hill [1], courtesy of Butterworths.

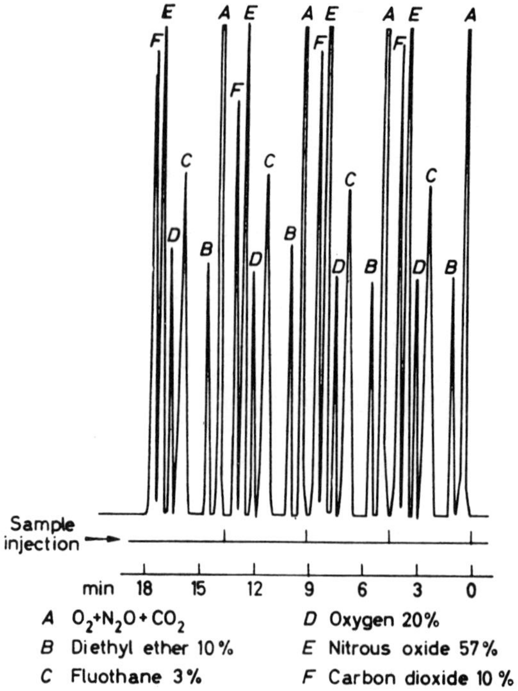

Figure 1.4. Routine monitoring of the output of an anesthetic machine. Dimethyl sulfoxide column in parallel with a dinonyl phthalate column. From Hill [1], courtesy of Butterworths.

Gases	Concentration (%)
Neon	0.1-0.30
Oxygen	20-95
Nitrogen	0-75
Carbon dioxide	0-10
Carbon monoxide	0.03-0.30
Acetylene	0.03-0.03

was performed without the necessity of temperature programming using a 4-ft x 1/2-in. SS column packed with 60-100 mesh silica gel for separation of acetylene and carbon dioxide and a 9-ft x 1/2-in. SS column packed with 40-60 mesh 5A molecular sieve for separation of neon, oxygen, nitrogen, and carbon monoxide. By appropriate positioning of a double four-way dual-column valve, the arrangement of the molecular sieve and silica gel columns is selective; that is, it may be either in series or in parallel as indicated in Figure 1.5. A typical gas chromatogram showing separation of neon, oxygen, nitrogen, carbon dioxide, carbon monoxide, and acetylene is shown in Figure 1.6.

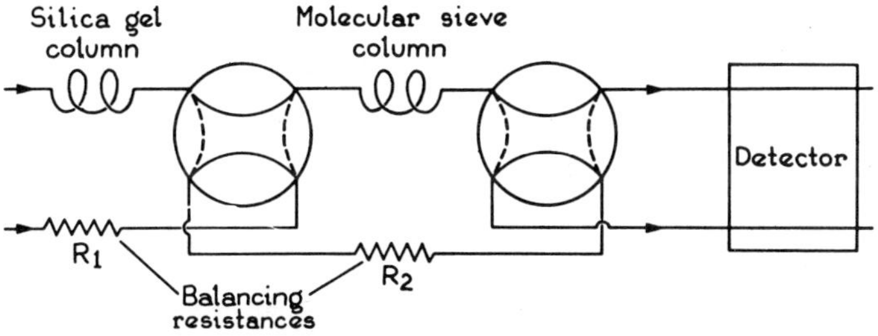

Figure 1.5. Column arrangement: 4-ft silica gel column and 9-ft molecular sieve column may be in series or parallel by appropriate positioning of double four-way dual-column valve. From De Graff [25], courtesy of Journal of Chromatographic Science.

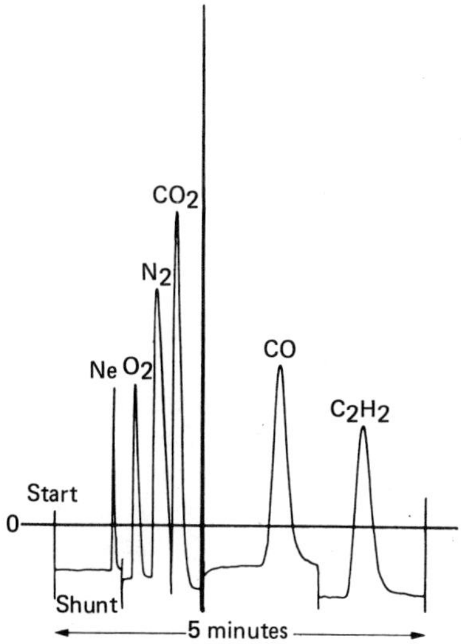

Figure 1.6. Gas chromatogram showing separation of neon, oxygen, nitrogen, carbon dioxide, carbon monoxide, and acetylene, as indicated. From De Graff [25], courtesy of Journal of Chromatographic Science.

As noted by De Graff [25], one of the major applications of gas chromatography in a cardiopulmonary laboratory is in the measurement of the diffusing capacity of the lung for carbon monoxide or, expressed another way, the transfer capacity of carbon monoxide from alveolar gas to intracorpuscular red cell hemoglobin. The apparent diffusing capacity of the lung for carbon monoxide may be defined as the transfer rate of CO from alveolar air to red cell hemoglobin per mm Hg alveolo-hemoglobin carbon monoxide gradient. Essentially a two-stage transfer process [(1) the passive diffusion of carbon monoxide across the alveolo-capillary membrane where gas transfer takes place between the small units of the lung and the blood vessels in contact with these lung areas, and (2) the chemical association or combination of CO with intracorpuscular hemoglobin], the pressure exerted by CO gas in combination with hemoglobin is negligible because of the strong chemical bonds formed between CO and hemoglobin. Under these conditions, the alveolo-capillary CO gradient is nearly the CO partial pressure within the alveoli. Consequently, the diffusion capacity calculation is rather simple, since only the CO partial pressure in the gas phase must be considered. As pointed out by De Graff, the apparent diffusing capacity is therefore equal to the instantaneous rate of disappearance of CO at any given time divided by the partial pressure of CO within the alveoli:

$$D_{LCO} = \frac{V_A}{P_{ACO}} \left[\frac{dF_{ACO}}{dt} \right] \tag{1.1}$$

where

D_{LCO} = apparent diffusing capacity of the lung for CO
V_A = alveolar volume
P_{ACO} = partial pressure of alveolar CO
F_{ACO} = fractional concentration of alveolar CO
t = time

The CO concentration at the start and end of breath holding is measurable, and if the differential equation is integrated, the diffusing capacity is defined by the change in fractional concentration of CO during breath holding. Since the rate of diffusion of CO across the alveolo-capillary membrane is predictably related to the O_2 diffusion rate across the same membrane boundary, the resistance to CO transfer offered by the alveolo-capillary membrane and by the chemical combination of CO with hemoglobin can be assessed independently of one another by varying the concentration of inspired oxygen (the CO-hemoglobin rate of combination varying with the partial pressure of oxygen in contact with red cell hemoglobin).

For example, since alveolo-capillary membrane diffusion for CO is directly proportional to that for O_2, disability resulting from impaired O_2 diffusion can be assessed from a knowledge of membrane-diffusing capacity for CO. In the measurement of diffusing capacity (Figs. 1.7 and 1.8) or lung blood flow, it is necessary to estimate the disappearance of one gas from the lung with reference to a second gas in the expired gas mixture which is not soluble in tissue or blood and hence does not disappear from the lung [25]. Consequently, any estimate of disappearance of reactive gas is therefore dependent on four separate gas analyses (that is, the concentration of reactive-nonreactive gases in the test mixture inspired and their concentrations in the expired gas mixture).

Because acetylene is a gas that is quite soluble and diffusible in blood, it is ideal for determining the measurement of lung capillary blood flow. As noted by De Graff, acetylene rapidly diffuses from alveolar air into the blood and traverses the capillaries lining the alveolar walls so that by the time blood leaves the capillary, virtual equilibrium has been established between the partial pressure of alveolar and blood acetylene. Since the rate of acetylene disappearance from alveolar air is a function of blood flow and Bunsen solubility of acetylene in blood (the solubility coefficient being constant and known), one is able to calculate blood flow. Since lung capillary blood flow is equal to cardiac output, its measurement is important in assessing disability of patients as a result of primary heart disease or heart disease secondary to lung disease.

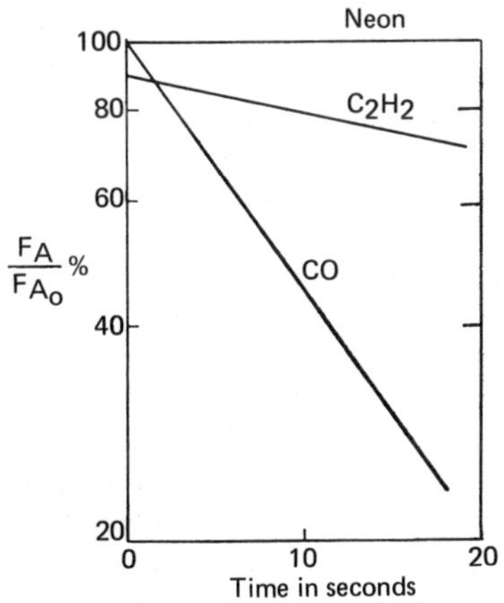

Figure 1.7. Change in lung carbon monoxide and acetylene concentration relative to neon during breath holding. Note depressed acetylene intercept indicating initial equilibration of acetylene in lung tissues. From De Graff [25], courtesy of Journal of Chromatographic Science.

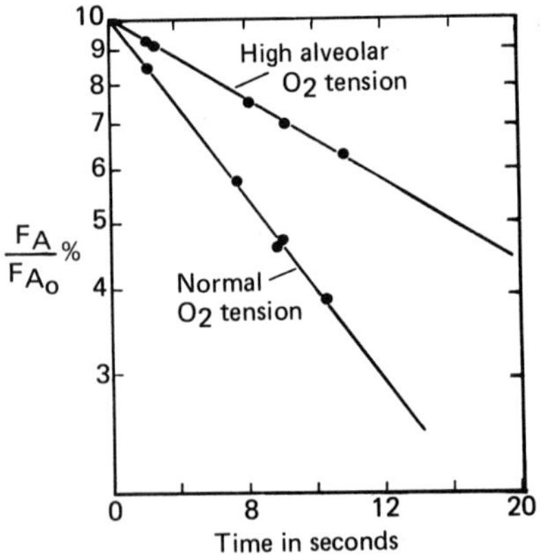

Figure 1.8. Disappearance of lung carbon monoxide relative to neon during breath holding in the presence of high and low lung partial pressures of oxygen. Membrane-diffusing capacity for carbon monoxide and lung capillary volume are calculated from the change in carbon monoxide disappearance rate resulting from altering oxygen concentration in the lung. From De Graff [25], courtesy of Journal of Chromatographic Science.

Jay et al. [26] and Hamilton and co-workers [27-29] also adapted the gas chromatographic technique for use in respiratory physiology. Jay et al. used the gas chromatograph as an analytical tool for measuring pulmonary functions, in contrast to the mass spectrometer and infrared analyzer used by Forster et al. [30] and Riley et al. [31] for determining the percentage composition of different mixtures of nitrogen, oxygen, carbon dioxide, helium, nitrous oxide, and carbon monoxide in the study of respiratory physiology. They noted that analytical problems are magnified spectrometrically if additional components, such as nitrous oxide or acetylene, are introduced into a gas mixture to measure pulmonary blood flow [32] and cardiac output [33]--especially when estimating pulmonary diffusing capacity [34,35] simultaneously with pulmonary capillary flow and volume. Using two sets of sample loops of different volumes to accommodate high and low gas concentrations with a Beckman GC-2 gas chromatograph, Jay et al. separated helium, oxygen, nitrogen, and carbon monoxide on a molecular sieve column prepared from copper tubing 12 ft x 1/2 in.-i.d. packed tightly with 20-30 mesh Linde 5A molecular sieves. Via a switching valve assembly, nitrous oxide and carbon dioxide were analyzed by a second column (an 18-in. length of stainless steel, 1/2-in.-i.d. tubing filed with equal 9-in. sections of 40-60 mesh charcoal and 20-30 mesh acid-washed firebrick). The area attributed to each component peak in the mixture was integrated with a recording integrator to relate area to percentage composition of the gas. The composition of gas mixtures containing chemically reactive components was analyzed by the method of Van Slyke and Sendroy [36] for comparison with the data obtained with the gas chromatograph. The results for a mixture of oxygen, nitrogen, and carbon dioxide used for this comparative study are shown in Table 1.1. In Figure 1.9, the results of the analyses with the Van Slyke apparatus are plotted against GC-derived data. However, in the Van Slyke method, corrections for argon were necessary since it was ordinarily included with the nitrogen in routine gas analyses for the calculation of the functional residual capacity.

For estimating cardiac output or pulmonary capillary blood flow and volume, the response of the GC instrument for the analysis of nitrogen, carbon dioxide, helium, and nitrous oxide was linear from 0 to 100%, as indicated in Figure 1.10. A typical chromatogram for CO_2 and N_2O analysis is shown in Figure 1.11.

On the other hand, Hamilton [27], in collaboration with the Fisher Scientific Co., developed a dual-column chromatograph that is particularly useful for respiratory gas analysis, as indicated in Figure 1.12. In this instrument, the sample is injected into the carrier gas stream and then carried through two chromatographic columns. Column 1 is 300 cm long and filled with a packing of 30% hexamethylphosphoramide on 30-60 mesh Chromosorb P, whereas column 2 is 250 cm long and filled with 40-60 mesh activated type 13X molecular sieve. A detecting element is located at the effluent end of each column (D1 and D2 in Fig. 1.12). Two reference elements (R1 and R2) are placed in the carrier gas stream. When the sample and carrier gas leave SV (sampling valve), they pass through a drying tube where moisture is removed before the sample undergoes

TABLE 1.1

Comparison of Data Obtained from Analysis of a Mixture of Oxygen, Nitrogen, and Carbon Dioxide Approximating Pulmonary Alveolar Air[a,b]

Method	N_2 + argon (mole %)	CO_2 (mole %)	O_2 (mole %)
Van Slyke and Sendroy	69.35±0.28	7.34±0.31	23.32±0.21
Gas chromatograph	69.36±0.34	7.30±0.10	23.30±0.19

[a]From Jay and Wilson [26], courtesy of Journal of Applied Physiology.
[b]Values are means ±S.D. There was no significant difference between the methods ($p < 0.01$).

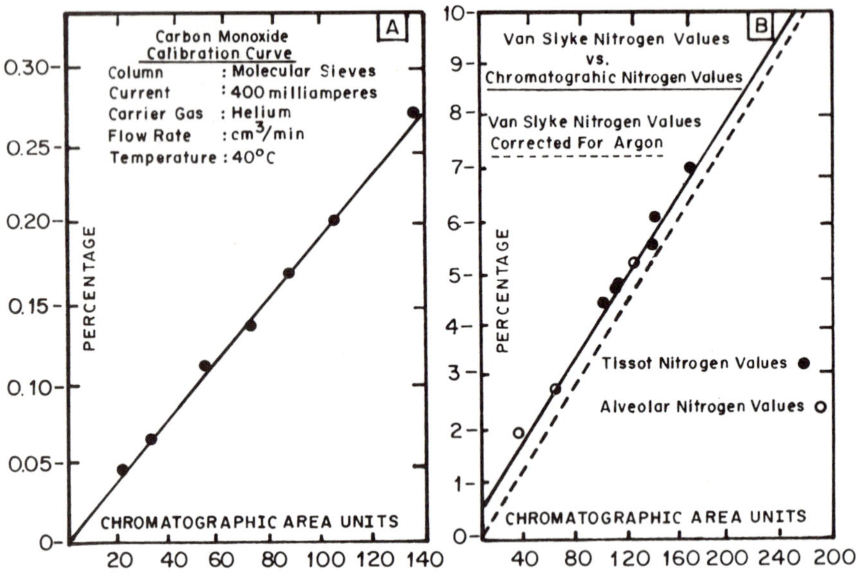

Figure 1.9. A. Calibration curve for trace amounts of carbon monoxide in air. B. Percentage of nitrogen in the alveolar and Tissot spirometer gases obtained during the pulmonary functional residual capacity measurements is plotted against units of chromatographic area. Dashed line is the same curve corrected for argon. From Jay and Wilson [26], courtesy of Journal of Applied Physiology.

separation. The gaseous sample enters the first column where CO_2 is retarded while the composite (O_2 and N_2) rapidly passes the first detector D1 and flows into the second column. As O_2 and N_2 pass through column 2, CO_2 is eluted from column 1 and detected by D1. In passing through column 2 (the molecular sieve), O_2 and N_2 are separated, eluted, and measured individually by detector D2; CO_2 is permanently adsorbed by the molecular sieve material so that it does not appear at D2. At a helium flow of 60 ml/min, a total analysis for CO_2, O_2, and N_2 is completed in 3.5 min (Fig. 1.13). With this system it is possible to observe differences between normal and abnormal conditions (emphysema), as indicated in Figure 1.14.

By substituting a 6-ft-long column filled with 30% DEHS (di-2-ethylhexyl sebacate) on 60-80 mesh Chromosorb P for column 1 in Figure 1.12, it is possible to resolve the gas mixture shown in Figure 1.15 using the model 29 Fisher-Hamilton gas partitioner. In measurements of lung diffusion capacity, column 1 is replaced by an empty tube. As noted previously, the patient breathes air containing a known ratio of CO or acetylene and neon.

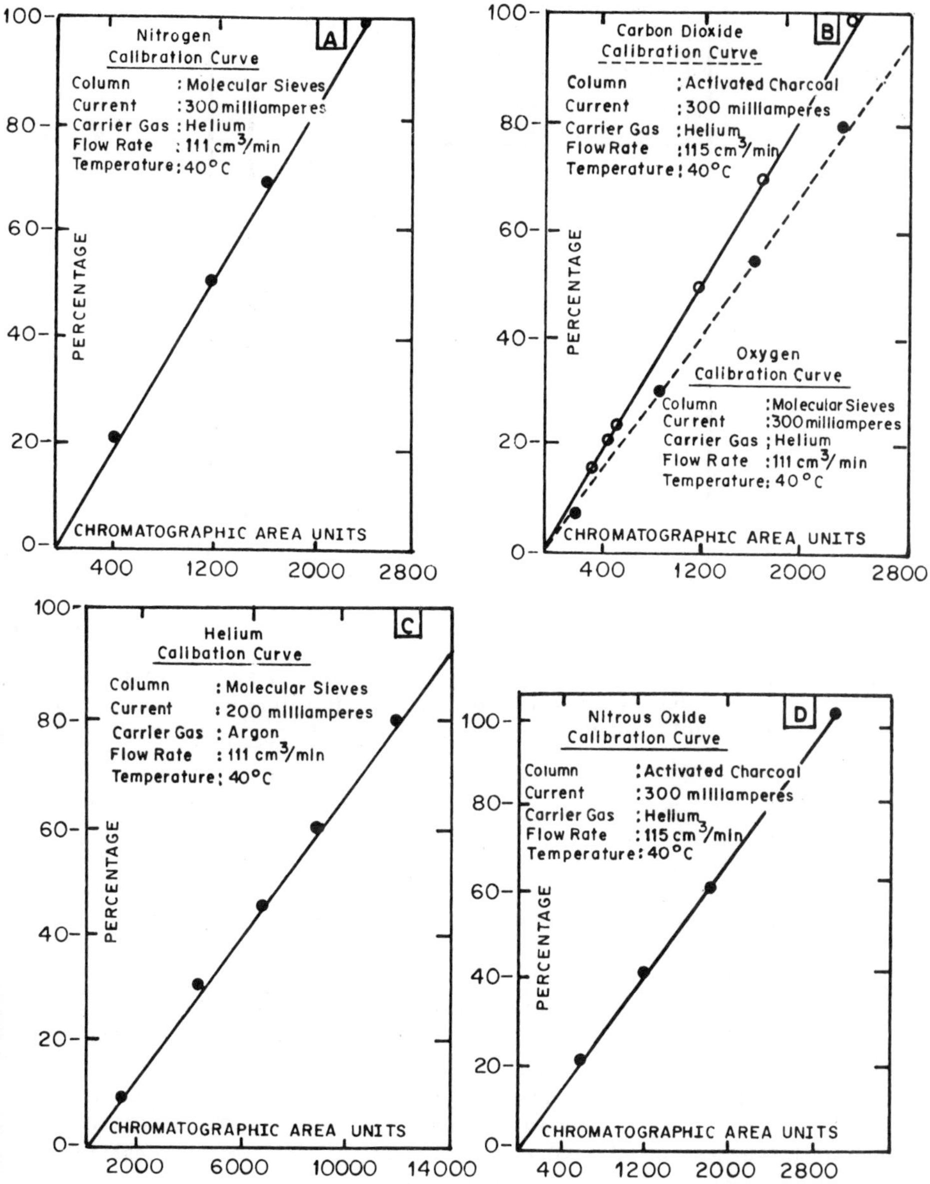

Figure 1.10. Calibration curves for nitrogen (A), carbon dioxide and oxygen (B), helium (C), and nitrous oxide (D). Percentage composition is plotted against units of chromatographic area. From Jay and Wilson [26], courtesy of Journal of Applied Physiology.

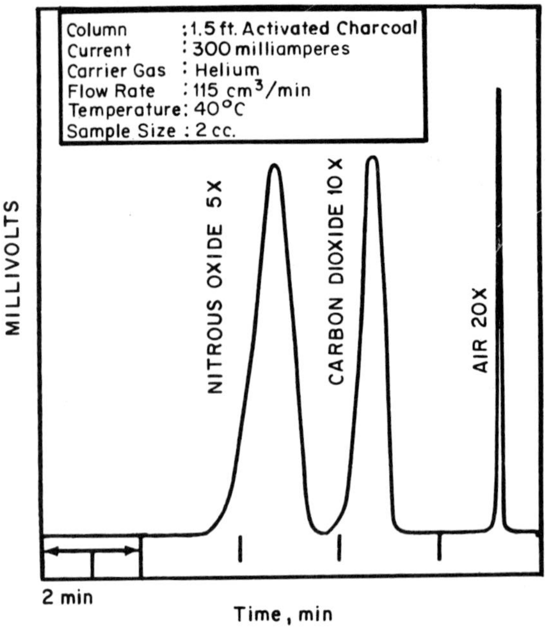

Figure 1.11. Tracing of nitrous oxide, carbon dioxide, and air analysis with integrator values printed in lower section. Baseline is stable. From Jay and Wilson [26], courtesy of Journal of Applied Physiology.

The patient holds his breath for a short time, during which the CO diffuses through the lung membranes. The neon is physiologically inert, so its concentration does not change. The air is then exhaled, and the CO/Ne ratio is measured and compared with the ratio in the test mixture. A typical chromatogram for such an analysis is illustrated in Figure 1.16.

A similar technique is used to measure lung capillary blood flow. In this case, acetylene is mixed with neon in the inspired air. The acetylene is highly soluble in blood and diffuses rapidly from alveolar air. The rate at which it disappears is correlated with lung capillary blood flow. The system used to obtain Figure 1.17 can be used to determine both diffusion capacity and capillary flow from a single test mixture containing CO, acetylene, and neon. To achieve this, column 1 in Figure 1.12 is replaced by a 12-in.-long composite column, packed with 6 in. of 28-200 mesh activated silica gel and 6 in. of Chromosorb P.

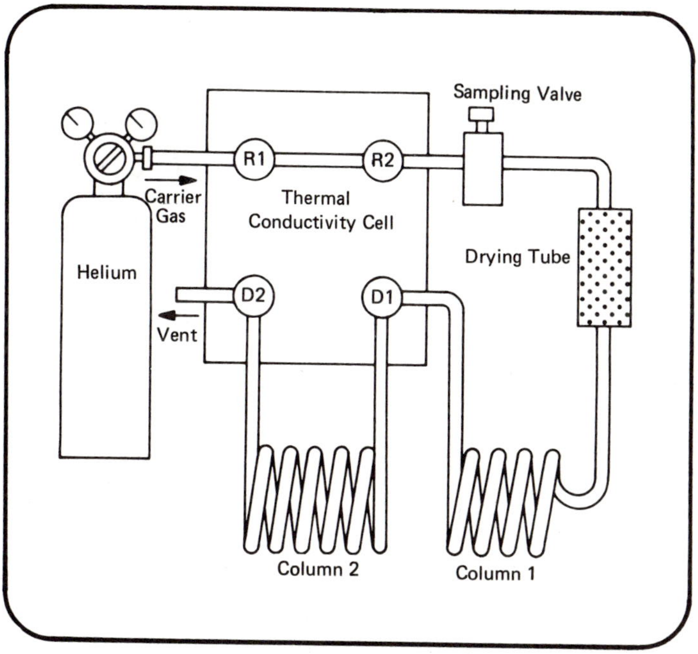

Figure 1.12. The dual-column/dual-detector chromatographic system used for respiratory gas analysis. Courtesy of Fisher Scientific Co.

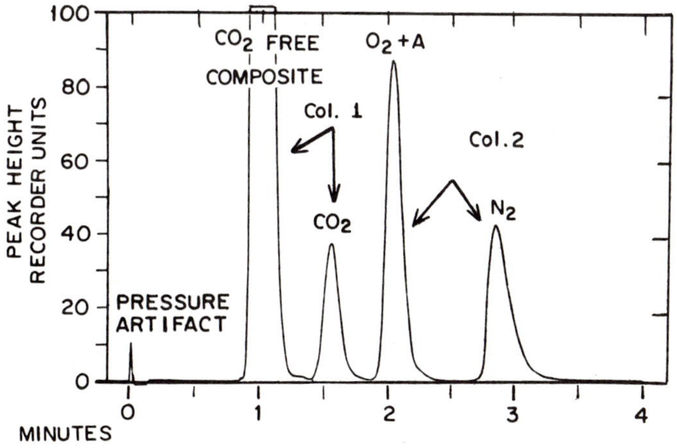

Figure 1.13. Chromatogram from a simulated respiratory gas sample analyzed with the dual-column, dual-detector gas chromatograph. Adapted from Hamilton [27].

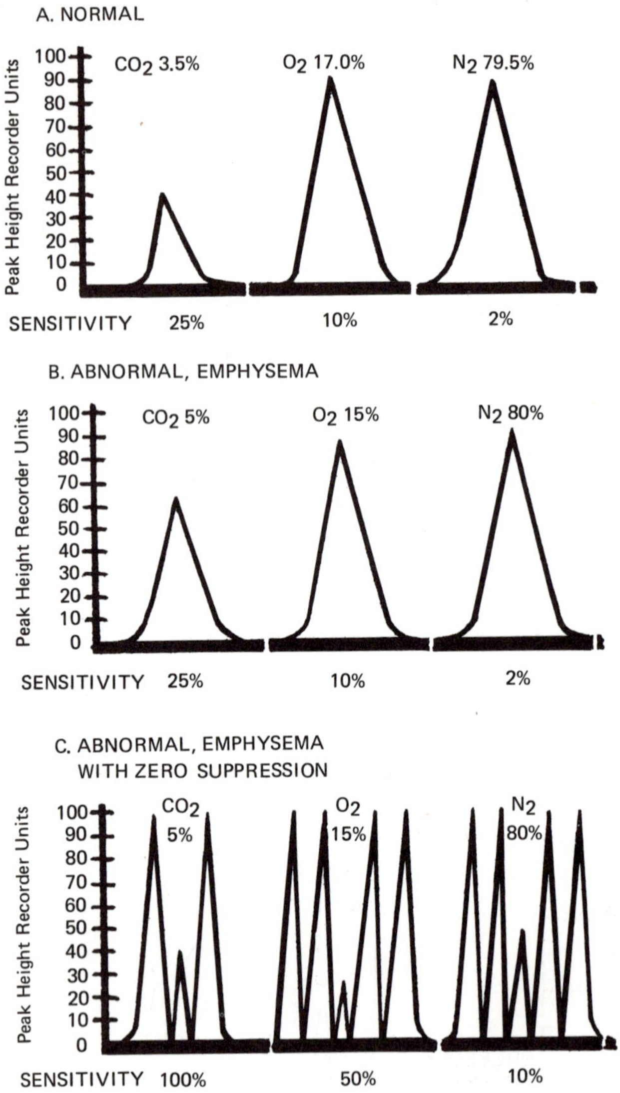

Figure 1.14. Tracings A and B compare CO_2 and O_2 in normal and abnormal conditions (emphysema). Both B and C are recordings of a patient with emphysema; C illustrates the increased accuracy of a recording by utilizing a zero suppression scale. Courtesy of Fisher Scientific Co.

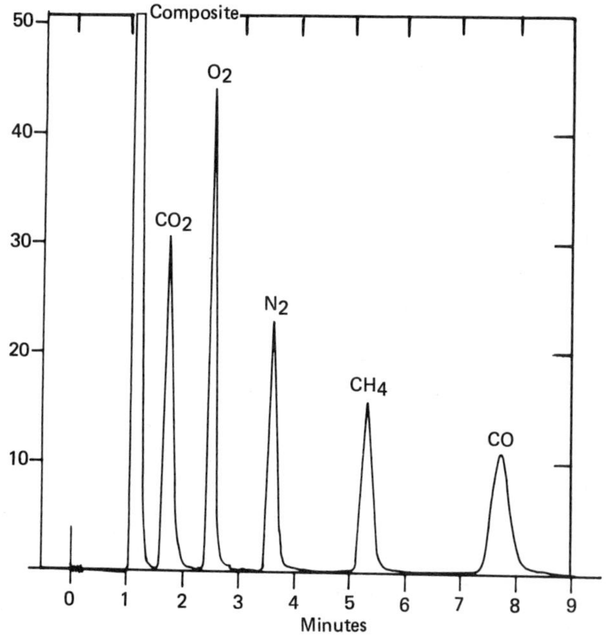

Figure 1.15. Typical separation of a five-component gas mixture. Courtesy of Fisher Scientific Co.

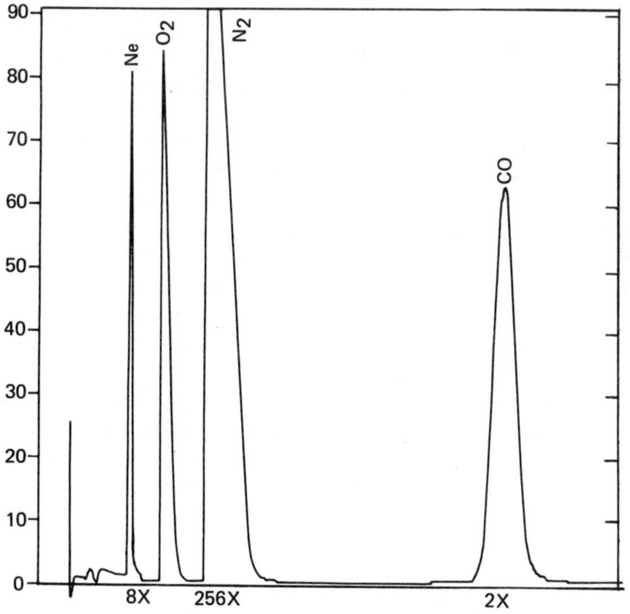

Figure 1.16. Determination of diffusion capacity by gas chromatography. Courtesy of Fisher Scientific Co.

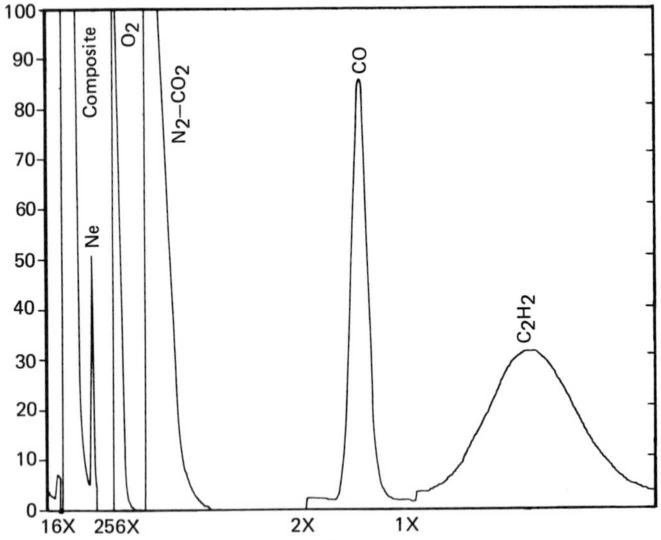

Figure 1.17. Measuring lung capillary blood flow. Courtesy of Fisher Scientific Co.

For the analysis of gases, a system such as that diagrammed in Figure 1.18 can be used where two columns, a porous polymer and a molecular sieve, are combined by a four-port rotary column switching valve. This arrangement results in all positive peaks without signal polarity switching, extremely long life for the molecular sieve, and ease of automation. The four-port valve allows carrier gas to flow through both columns to the detector, or only through the porous polymer column to the detector. Figure 1.19 shows a typical chromatogram achieved with this system, where temperature programming was employed to speed the analysis of the higher hydrocarbons eluting from the porous polymer column. Using the system shown in Figure 1.18, the sample is introduced with the two columns in series and the composite peak goes directly into the molecular sieve column without passing through the detector. Hammerstrand [37] then rotates the switching valve to dead-end the molecular sieve column and the CO_2 is eluted from the porous polymer column. The valve is then switched back to provide flow through the molecular sieve column and the O_2 and N_2 elute through the detector. By continuing this switching process, as shown in Figure 1.19, all components can be separated and detected in a relatively short time. The time can be significantly reduced by using a temperature program as shown.

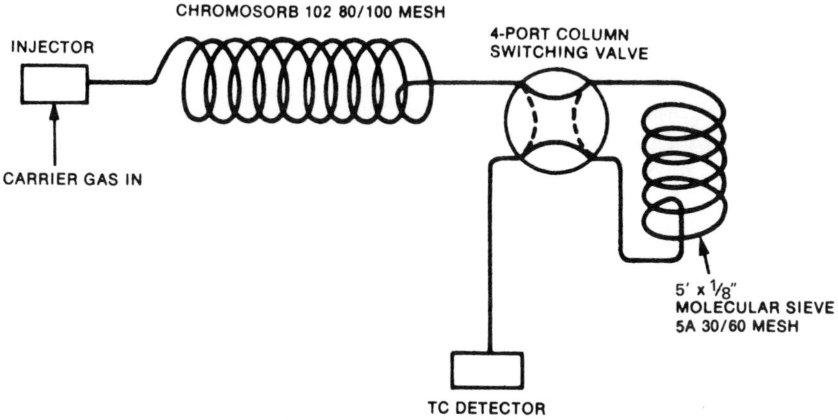

Figure 1.18. Dual-columns system, with switching valve. From Hammerstrand [37], courtesy of Varian Associates.

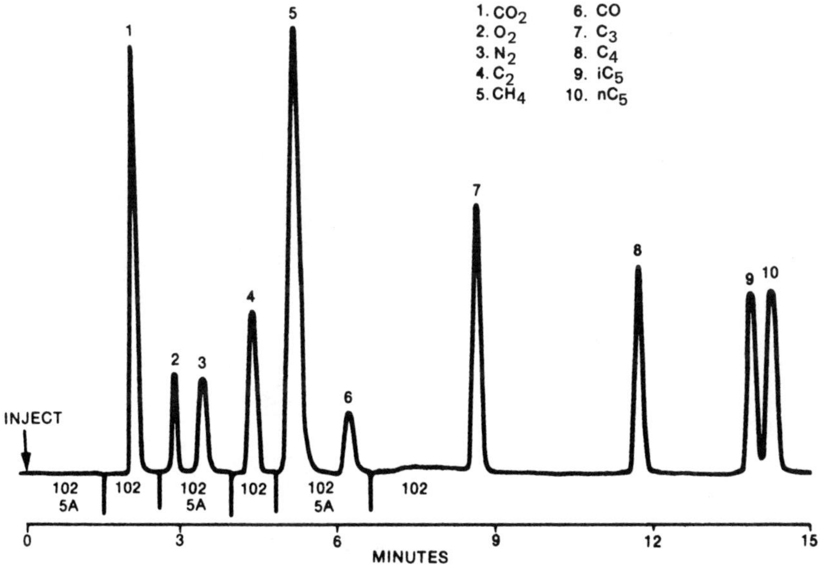

Figure 1.19. Gas analysis conditions: TCD at 200°C; dual columns, 10 ft x 1/8-in. SS with 80/100 Chromosorb 102 and 5-ft x 1/8-in. SS with 30/60 molecular sieve 5A. Temperature program: isothermal at 65°C, then program at 15°C/min from 65 to 200°C. He at 30 ml/min. From Hammerstrand [37], courtesy of Varian Associates.

Patzelova [38] used a two-column tandem arrangement as illustrated in Figure 1.20. The first glass column (50 cm x 4 mm i.d.) was packed with Chromaton N-AW impregnated with 15% Kel-F oil 10, whereas the second column (270 cm x 4 mm i.d.) was filled with 100-120 mesh Poropak Q. Using a H_2 flow rate of 70.6 ml/min, the analysis of halothane, O_2, CO_2, and N_2O was performed in 6 min at room temperature with the detector maintained at 40°C.

Since Hollis [39] first proposed the use of porous hydrocarbon polymers as the sole packing for gas chromatographic separations, various investigators have reported on the GC characteristics of synthetic polymer beads [40-47]. Recently, Fritz and Chang [48] investigated the "XAD" series of macroporous resins for sorption and chromatographic separation of gases.

Fritz and Chang separated several sample mixtures to show the practicality of gas chromatography with an XAD resin as the column packing. For example, on an XAD-4 column, a mixture containing four rare gases, methane, and oxygen was completely resolved (Fig. 1.21). This separation was achieved on a 2.5-ft column, although a 4-ft column gave a slightly sharper separation. The temperature program was 4.0 min at ambient temperature, followed by a 50°C/min temperature increase to 80°C, and a constant 80°C thereafter.

On the other hand, the analysis of inert and chemically active biologic and experimental gases in solution in the body fluids of experimental animals and human beings has also been of great concern to many investigators [26,27,49-64]. Gas chromatographic techniques have been developed which are applicable to any chemically combined mixture of dissolved gases in blood and body fluids of animals, sea water, or any other gas in solution, provided that their concentrations are sufficient for detection. For example, Wilson and co-workers [26,49,51,65], in addition to their use of the gas chromatograph as an analytical tool for measuring the composition

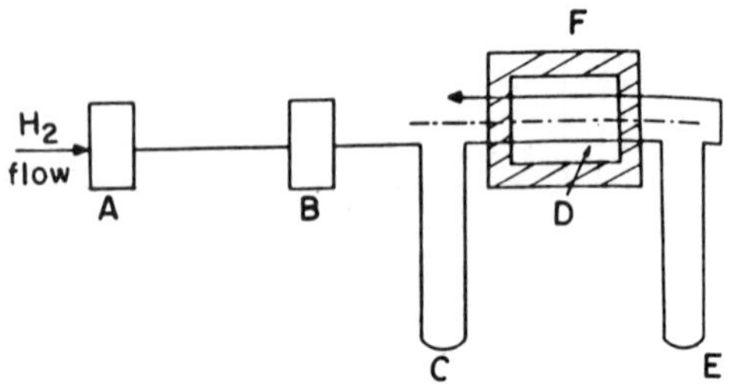

Figure 1.20. The function scheme of the chromatograph used

A = needle microvalve
B = sampling valve
C = first column
D = thermal conductivity detector
E = second column
F = thermostatic oven of the detector

Adapter from Patzelova [38].

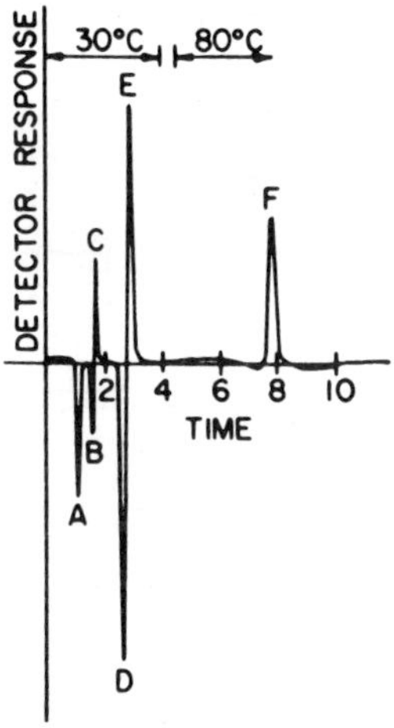

Figure 1.21. Gas chromatographic separation of rare gases. Conditions: 4.0-ft. x 1/4-in. column, 100-150 mesh XAD-4, flow rate 20 ml/min (N_2), thermal conductivity detector. Retention time: (A) neon, 1.1 min; (B) oxygen, 1.6 min; (C) argon, 1.8 min; (D) methane, 2.7 min; (E) krypton, 2.9 min; and (F) xenon, 7.8 min. From Fritz and Chang [48], courtesy of Analytical Chemistry.

of respiratory gases [49], developed a GC method and apparatus for analysis of blood gases [26] as well as dissolved inert and chemically combined gases with hemoglobin and bicarbonate in blood samples as small as 0.25 ml [65].

In these related studies, a Beckman GC-2 gas chromatograph was modified as shown in Figure 1.22, which shows the direction of helium flow through the apparatus and an analytic cuvette constructed from a Pyrex tube with an outside diameter of 25 mm and a fritted glass filter disk 20 mm in diameter having a pore size ranging from 40 to 60 μm (Fig. 1.23). A one-way high-vacuum Pyrex glass stopcock was sealed to the cuvette (6 cm in height) just above the fritted glass disk for removal of the blood sample and exhausted reagents. For admission of the blood, a 0.5-ml graduated pipette, calibrated in 0.01-ml divisions, was sealed to the first arm of the Pyrex glass three-way stopcock, which had a 100-120 capillary bore. The second arm of the stopcock was attached to the mixing chamber and the third arm of the stopcock was used for loading the pipette with blood, loading the chamber with reagents, and cleaning the apparatus after the analysis of each sample [49,51].

The second arm of the stopcock, which was to be attached to the analytic chamber, was extended to a fine point and bent to a right angle to facilitate the delivery of a known quantity of blood from each blood sample analyzed.

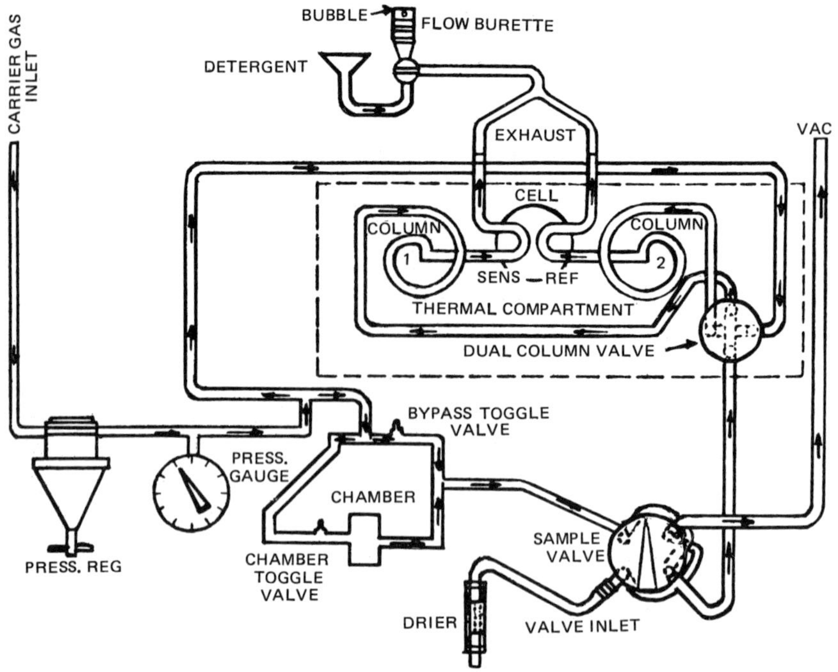

Figure 1.22. Schematic diagram of Beckman GC-2 apparatus modified for blood gas analysis. From Wilson [49], courtesy of Journal of Applied Physiology.

The dead space in this section of the sidearm was determined and subtracted as a correction factor from each sample of blood. The right angle, with the tip pointing downward toward the analytic solution, prevented sample loss on the chamber's walls. Without increasing or decreasing the carrier gas pressure in the GC system, the apparatus was designed so as to permit pressure--compressed air, vacuum, or atmospheric pressure --to be exerted on the measuring pipette and loading arm of the stopcock to facilitate sample injection into and cleaning of the apparatus. Sample introduction into the chamber was accomplished by impressing a pressure above the sample in the measuring pipette greater than that exerted by the carrier gas, a force sufficient to push the sample into the analytic Pyrex glass cuvette. A known volume of sample ranging from 0.25 to 0.35 ml was admitted into the chamber containing 2 ml of a 1:1 solution of saponin and potassium ferricyanide (16 g of saponin and 64 g of potassium ferricyanide per liter of distilled water) and 1 ml of 0.5 N lactic acid, to which was added 0.2 ml of Antifoam 60. Since the lactic acid was present

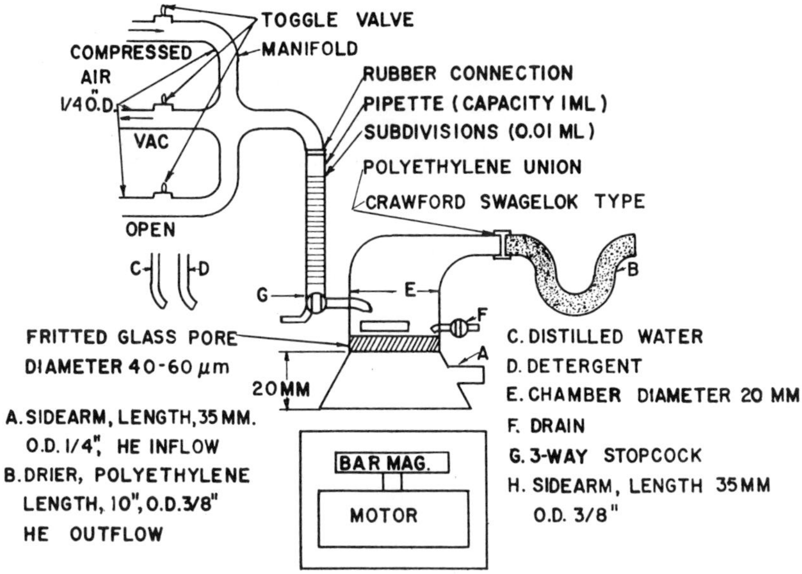

Figure 1.23. A schematic design of the Pyrex glass cuvette, manifold, magnetic stirring bar, and drier. O.D. refers to outside diameter; HE to helium or any other carrier gas. From Wilson [51], courtesy of Journal of Chromatographic Science.

in the solution to liberate the carbon dioxide bound in the form of sodium bicarbonate, the chamber should have contained 2 ml of saponin and potassium ferricyanide plus 0.2 ml of Antifoam 60 in the analysis of blood oxygen. To enhance the rapid mixing of the blood and analytical reagents (and, in turn, the liberation of the oxygen from the hemoglobin) in the chamber, a small magnetic stirring bar was placed in the solution as indicated in Figure 1.23 [49,51]. The deaeration procedure was performed by passing helium through the solution in the chamber until the chromatographic curve returned to baseline.

The blood sample was delivered rapidly into the chamber after the initial reading of its level in the pipette has been noted. In approximately 3 to 4 min after the gas chromatograph had analyzed the gas mixture released from the blood, a final blood level reading was recorded; this short time lag was sufficient to permit the blood clinging to the pipette's wall to drain down to the remaining blood level. After each sample was analyzed, both the pipette and stopcock were washed with detergent, rinsed with distilled water, and dried. The chamber was washed with distilled water following the analysis of two samples when analytical reagents were changed.

Following the release of the physiological gases from the blood, the helium carrier gas sweeps these liberated gases through the polyethylene tubing containing a drying agent. For the analysis of oxygen and carbon dioxide, two columns were used: one was 5 ft long and filled with 30-60 mesh molecular sieve for the resolution of nitrogen and oxygen (CO_2 being irreversibly absorbed with this packing), and the other was 18 in. in length, packed with activated charcoal which separated the carbon dioxide from both the nitrogen and oxygen.

The carbon dioxide and oxygen concentrations, expressed in milliliters per 100 ml of blood, was determined as shown below:

$$C_{BO_2} = \frac{V_L}{V_B} \frac{F_{SO_2}}{A_S} (A_U) \tag{1.2}$$

where

C_{BO_2} = O_2 content in blood (vol. %)
C_{BCO_2} = CO_2 content in blood (vol. %)
V_L = volume of sampling loop in the gas chromatograph corrected to standard temperature and pressure (dry)
V_B = volume of blood injected into the analytical chamber
F_{SO_2} = known volume percent of O_2 used in the calibration procedure of the standard gas sample
A_S = area under the curve obtained when standard containing known content of either CO_2 or O_2 was analyzed
A_U = area under the curve obtained when the blood sample was analyzed for unknown content, either CO_2 or O_2

The CO_2 content in blood was readily determined by substituting C_{BCO_2} and F_{SCO_2} in Eq. (1.2). As outlined by Jay and Wilson [26], the apparatus was calibrated with a gas mixture of known composition, whereas a sodium bicarbonate solution of known molarity was used for CO_2 calibration.

Hamilton [27] and Galla and Ottenstein [50] used a Fisher gas partitioner modified to accommodate a reaction chamber as shown in Figures 1.24 and 1.25 for the analysis of blood samples containing O_2 and CO_2 and acetylene, cyclopropane, ethylene, and nitrous oxide, respectively. In their dual-column, dual-detector GC system which consisted of a reaction chamber, column system, detector unit, and recorder, the carrier gas (helium) passed two references detectors prior to entering the reaction chamber through a buffer tube filled with an inactive porous material (Chromosorb P). In the arrangement used by Galla and Ottenstein, reagents or blood injected through the sample inlet of the reaction chamber were stirred by a magnetic stirrer and the helium swept the released gases from the chamber through a drying tube prior to their passage through

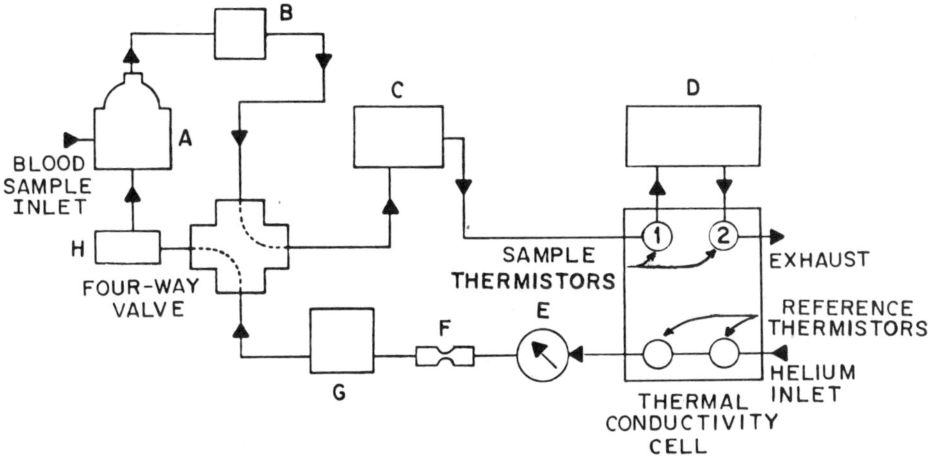

Figure 1.24. Schematic diagram of chromatographic system. A, reaction chamber; B, drying tube; C, column 1; D, column 2; E, pressure gauge; F, orifice; G, gas sample inlet; H, buffer tube. Adapted from Galla and Ottènstein [50].

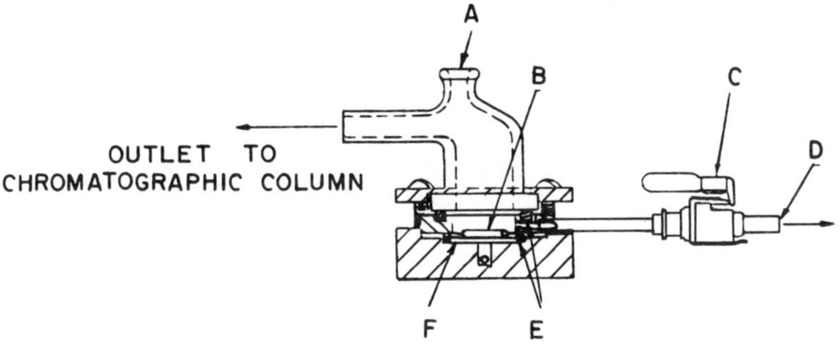

Figure 1.25. Schematic diagram of reaction chamber. A, sample inlet; B, magnet; C, valve; D, drain; E, "O" rings; F, stainless steel mesh. Adapted from Galla and Ottenstein [50].

column 1 (consisting of a 20- x 1/4-in. aluminum tube packed with 60-80 mesh Chromosorb P) which subsequently led to the first sample detector. Upon elution from column 1, the gases were recorded by the first detector as a composite peak. The components of the gas mixture were resolved by column 2 and upon elution were seen by the second detector. For the separation of oxygen and carbon dioxide in blood, column 2 was packed with silica gel. Table 1.2 lists the columns most suitable for analyzing other inert gases that can be suitably resolved from nitrogen. The reagent charge, consisting of 0.2 to 0.4 ml of octyl alcohol and 0.5 ml of potassium pyrogallate solution, was injected into the chamber and allowed to deaerate for several minutes. Blood samples were collected anaerobically from the tonometers in hypodermic syringes sealed with rubber septum caps. Using a special Hamilton microliter syringe with a Chaney adapter for precise measurement, a portion of the blood sample in the sealed hypodermic syringe was removed and injected carefully into the reaction chamber. For the gases tested, peak area measurements were proportional to the gas present in the sample. On their respective columns (Table 1.2), the retention times for acetylene, cyclopropane, ethylene, and nitrous oxide were 4.0, 5.5, 3.25, and 3.0 min, respectively. As noted in Table 1.3, satisfactory reproducibility was achieved; the magnitude of the coefficient of variation for each analysis related to the sharpness of the peak and the degree of separation.

TABLE 1.2

Column Systems for Analysis of Inert Gas[a]

Component	Column packing	Column dimensions
Acetylene	30% bis[2-(2-methoxyethoxy)ethyl] ether on 60/80 mesh Chromosorb P	6 ft x 1/4 in.
Cyclopropane	Same as above	12 ft x 1/4 in.
Ethylene	Silica gel, 60/200 mesh, Davidson type, Grade no. 590	12 in. x 3/16 in.
Nitrous oxide	30% bis[2-(2-methoxyethoxy)ethyl] ether on 60/80 mesh Chromosorb P	12 ft x 1/4 in.
	30% hexamethylphosphoramide on 60/80 mesh Chromosorb P	12 ft x 1/4 in.

[a]Adapted from Galla and Ottenstein [50].

TABLE 1.3

Reproducibility of Analyses on Triplicate Aliquots[a]

Gas	Coefficient of variation (%)
Acetylene	±1.5
Cyclopropane	±2.9
Ethylene	±6.0
Nitrous oxide	±2.1

[a]Adapted from Galla and Ottenstein [50].

Using a plot of chromatographic areas versus the inert gas partial pressure, fairly accurate volume percent values could be obtained by multiplying the area (expressed as partial pressure) by the Ostwald solubility coefficient. However, the system did have one disadvantage; the delayed release of the gases from the blood in this continuous flow-through apparatus and the dead space of the reaction chamber resulted in a broadening of the chromatographic peaks, necessitating the use of longer columns for separation of the components than would have been needed with a more instantaneous release.

In contrast to the continuous flow-through arrangement used by Galla and Ottenstein, Hamilton [27] used a system based on equilibrating blood with helium, after which the gas mixture was flushed into the chromatograph (Fig. 1.26). The reaction chamber was similar to that used by Galla and Ottenstein, but it was interposed in the carrier gas stream with a bypass valve that permitted the chamber to be isolated from the flow of helium. Removal of the gases was achieved by equilibrating the blood sample with helium and subsequently washing the gases into the main carrier gas stream by changing the valve position. In this mode, the carrier gas flowed over the mixture and swept the extracted gas into the GC unit in a reproducible way each time. Since the sample introduction pattern was reproducible, peak height measurements were used. It should be noted, however, that the reaction chamber was temperature regulated, so a constant proportion of the released gas remained in solution.

Hamilton compared the helium equilibration technique for extracting blood gases with an improved Fisher helium flow-through chamber and the manometric Van Slyke method for measuring whole-blood CO_2 content. Both chromatographic systems (that is, the equilibration and flow-through methods) demonstrated a linear response with increasing CO_2 concentration.

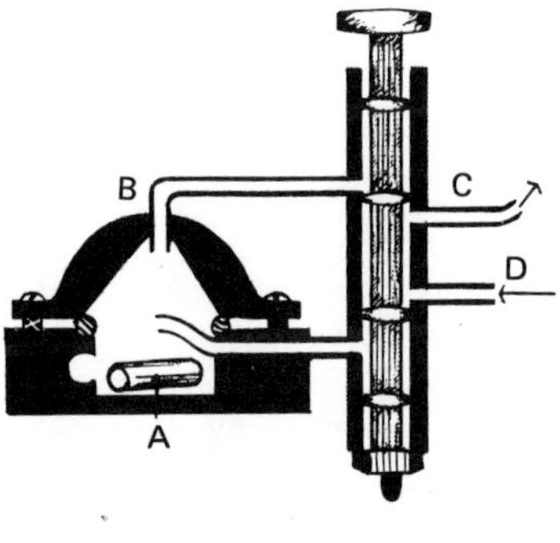

Figure 1.26. Helium equilibration chamber developed for the extraction of blood gases and their subsequent introduction into the gas chromatograph for analysis. A, stirring magnet; B, equilibration chamber; C, bypass valve; D, helium flow. Adapted from Hamilton [27].

The flow-through system provided a mean difference of 0.27 ± 0.72 mEq of CO_2/liter when compared with the Van Slyke analysis. The difference between the values for the equilibration system and the Van Slyke system was 0.32 ± 0.30 mEq of CO_2/liter. In both cases the mean Van Slyke value was higher than that obtained with the gas chromatograph, but the differences were not statistically significant.

In 1965, Yee [52] described a practical procedure for calibrating a chromatographic system for the determination of blood gases. Stoichiometric amounts of carbon dioxide and oxygen were generated in the reaction chamber from potassium carbonate and hydrogen peroxide, respectively. Other means of calibrating blood gas measuring systems have used pure gas mixtures [57], oxygen-saturated blood [58], and sodium carbonate [66]. For this study, a Beckman GC-2A gas chromatograph with blood gas accessory was fitted with columns of 6-ft x 1/4-in. Linde 5A molecular sieve, 4-ft x 3/16-in. silica gel, and 8.5-ft x 1/4-in. dry firebrick. The extraction chamber consisted of a glass tube (Beckman 94276) fitted with a frit made of Teflon, a small bar magnet, and a serum cap. A statistical evaluation (F and t tests) of the O_2 and CO_2 (Table 1.4) data indicated no significant differences between results obtained by either gas chromatography (GC) or Van Slyke (VS) procedures.

TABLE 1.4

CO_2 Content of Arterial and Venous Blood Specimens[a]

	Vol./100 ml found			
Sample no.	GC	VS	VS, corr.[b]	Rel. diff.
1	41.2	39.7	41.2	0.0
2	39.8	37.4	38.8	+1.0
3	36.5	35.8	37.2	-0.7
4	43.5	41.3	43.3	+0.2
5	65.5	62.5	65.5	0.0
6	44.4	42.9	44.3	+0.1
7	39.9	39.4	40.7	-0.8
8	36.6	35.4	36.6	0.0
9	35.8	34.3	35.2	+0.6
10	35.3	34.1	35.1	+0.2
±S.D.	0.66	0.68		

[a]From Yee [52], courtesy of Analytical Chemistry.

[b]F and t were not significant at a 95% confidence level.

Other investigators, such as Farhi et al. [53], Ikels [54], and Natelson and Stellate [55], have used similar GC approaches for the determination of nitrogen dissolved in blood.

Farhi et al. determined dissolved nitrogen in blood by gas chromatography and (a - A) nitrogen difference. By combining vacuum extraction in a Van Slyke chamber and separation of the extracted gases in a gas chromatograph, it is possible to determine the nitrogen (N_2) content of 1.5 ml of blood or other biological fluids in less than 10 min. The 95% confidence limits are 0.44% on either side of the mean of the triplicate analysis, or 2.4 mm P_{N_2} in arterial blood when breathing room air. Application of the method to the problem of arterial (a) - alveolar (A) N_2 difference yields the following data: (1) N_2 solubility in whole blood at 37.3°C varies from 0.0125 to 0.0129; (2) N_2 solubility in urine is inversely related to urine specific gravity, confirming Klocke and Rahn's data [67]; (3) changes in arterial N_2 content are reflected in arm superficial venous blood and urine N_2 only after a considerable period of time, indicating that either of these will give an excellent indication of the mean P_{N_2} over a period of time; (4) there is no systematic difference between venous blood and urine P_{N_2}; and (5) the (a - A) N_2 difference in nine normal subjects varied from 3.7 to 13.1 mm Hg. As noted by the authors, the method consists of a single vacuum extraction in a modified Van Slyke apparatus and separation of the evolved gas in a gas chromatograph. A complete analysis requires less than 10 min, and with practice, ten measurements can be obtained in 1 hr.

Applied to the determination of the Bunsen solubility coefficients of nitrogen in distilled water at 37, 32, 27, and 3°C, the results obtained compare favorably with values listed in international critical tables as indicated in Table 1.5. These data are also plotted in Figure 1.27. The success of the initial attempts to determine N_2 solubility in distilled water was extended to blood. The values found were of the same order as those reported for ox blood and plasma by Van Slyke et al. [68] (0.0130 and 0.0117 for blood and plasma, respectively). Farhi et al. calculated a mean of 0.01277 for nine subjects at 37.3°C.

Although the method proposed by Ikels [54] incorporated similar apparatus as that used by Farhi et al. [53], several significant alterations were required in order to measure accurately very small volumes of N_2 dissolved in blood. In the technique suggested by Ikels, a modified Van Slyke apparatus and a gas chromatograph are used as a single instrument for quantifying small volumes of N_2 dissolved in blood. The principle of the method consists of first extracting the inert and chemically reactive gases, followed by the absorption of the O_2 and CO_2 and, finally, measuring the inert gases by gas chromatography. As pointed out by Ikels, his method was devised primarily for the analysis of dissolved N_2 in blood during periods of breathing "pure" O_2 or the diverse mixtures of O_2 that may be used as interior gases in manned space vehicles.

TABLE 1.5

Bunsen Solubility Coefficients (α) for Nitrogen in Distilled Water[a]

Temperature (°C)	α N_2 in water Present data	Int. crit. tables
3.10	0.02218	0.02178
26.80	0.01430	0.01432
27.15	0.01425	0.01427
32.00	0.01337	0.01345
32.05	0.01336	0.01344
37.05	0.01270	0.01272

[a]From Farhi et al. [53], courtesy of Journal of Applied Physiology.

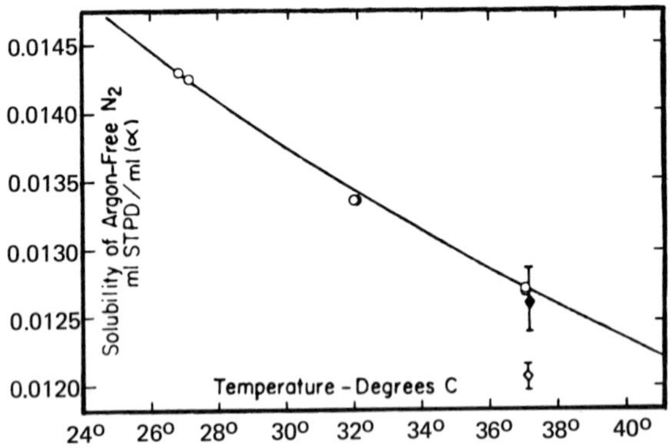

Figure 1.27. Relationship between solubility of nitrogen and temperature. Circles are experimental data for distilled water. Continuous line is plotted from data in the literature. Black diamond is data obtained in fresh human blood (mean and S.D.) and white diamond represents data obtained on human plasma. From Farhi et al. [53], courtesy of Journal of Applied Physiology.

A comparison of the results for Bunsen absorption coefficients by the method proposed by Ikels, in which the chemically active biologic gases were first absorbed prior to injection into the GC, indicates excellent agreement with the values for N_2 in water and, of greater importance, the values for blood compare favorably with the mean value reported by Farhi et al. Ikels concluded that these results lend support to the validity of the method. The method is accurate to within 0.3 μl of N_2 gas and is suitable for use throughout the physiological range of dissolved N_2 which may be encountered in nitrogen elimination studies. The effect of O_2 breathing on N_2 content of the antecubital venous blood showed that approximately 28% of the original N_2 content was present after 0.5 hr, 14% after 1 hr, 10% after 2 hr, and 8% after 4 hr.

To avoid carrier gas flow interruption and incomplete removal of the volatile gas being analyzed from the solvent or biological fluid, Natelson and Stellate [55] designed an apparatus for extraction of gases which could be subsequently injected into the gas chromatograph. The principle used is to introduce the specimen into a microgasometer (Fig. 1.28) [69]. After release of the gas under vacuum, the solvent is sequestered in a storage chamber, and the gas is injected into a flowing stream of carrier gas. The gas chromatograph used was the Fisher Clinical Gas Partitioner, with constant-temperature stabilizer. Equipped with thermistor detectors and a silica gel column followed by a molecular sieve column, the reaction chamber was removed from the instrument and the helium flow pattern modified to permit uninterrupted flow through the chamber attached to the microgasometer. Operational details of the instrument and analytical procedure (covering such parameters as sampling, release of gases and sequestering of reagents, injection of the gas into the chromatograph, preparing the instrument for the succeeding sample, and preparation of standards for calibration) are carefully discussed by the authors in the text. The instrument as illustrated in Figure 1.28 is a schematic representation of its total arrangement, where one notes that the injection chamber consists of a glass tube with a capillary sealed into it. Although not shown, ball joints at the entrance and exit of the flow chamber permit its easy connection and removal for cleaning from the source of gas and the chromatograph. In this proposed analytical technique, the shape of the peak is influenced by the shape and capacity of this chamber; sharper peaks are obtained as the capacity of the chamber is reduced. Since sampling is piston controlled, it is rapid and accurate; the total time needed for gas release and injection is usually less than that required for the development of the chromatogram. Since the shapes of the curves are reproducible, peak height measurements are used for assay. Comparative results using this equipment are listed in Table 1.6 for eight consecutive blood analyses for oxygen by gasometer and chromatography compared to the iron content determined by X-ray spectrometry [70,71]. For comparison, results are

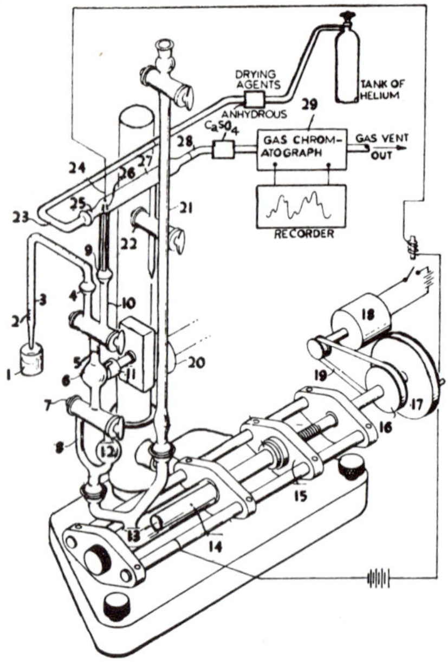

1. Container with sample
2. Graduations
3. Bent pipette
4. Ball joint
5. 0.12-ml mark
6. Reaction chamber
7. Lower two-way stopcock
8. Bypass tube
9. Ball joint
10. Connecting tube
11. Rotating magnet
12. Reservoir
13. Glass barrel
14. Piston
15. Movable guide
16. Pulley wheel
17. Hand pulley
18. Motor
19. Belt
20. Motor
21. Manometer
22. Stopcock for drain
23. Connecting tube
24. Platinum contact
25. Ball joint
26. Injection capillary
27. Flow chamber (8-mm bore)
28. Connecting tube to chromatograph
29. Gas chromatograph

Figure 1.28. Gas extractor and injector. From Natelson and Stellate [55], courtesy of Analytical Chemistry.

TABLE 1.6

Comparison of Values Obtained for Oxygen in Whole Blood by Microgasometer and Gas Chromatograph[a,b]

Blood no.	Gasometer		Chromatograph		X-ray[c]	
	Vol. %	Hemoglobin (g/100 ml)	Vol. %	Hemoglobin (g/100 ml)	Hemoglobin (g/100 ml)	Hematocrit (%)
1	10.45	7.78	10.22	7.62	7.85	22
2	10.88	8.12	10.02	7.60	8.01	23
3	26.10	19.47	27.40	20.40	19.13	56
4	18.00	13.43	18.86	14.07	13.80	40
5	18.78	14.00	18.83	14.04	14.20	42
6	18.06	13.47	17.73	13.00	13.10	40
7	18.48	13.78	17.64	13.01	13.02	40
8	20.93	15.61	20.29	15.14	15.01	43

[a]From Natelson and Stellate [55], courtesy of Analytical Chemistry.

[b]Values are mean of triplicates in each case. Factor for conversion of volume percent to hemoglobin value was 0.746.

[c]Calculated from iron content, 50 mg/100 ml of Fe = 14.7 g of hemoglobin. Standard deviation (precision) for the values in volume percent for gasometer ±0.21, for chromatograph ±0.25. For hemoglobin by X-ray ±0.12.

converted to grams of hemoglobin. From the data, Natelson and Stellate concluded that no significant difference is observed among the three methods when the t test is applied for significance of mean differences. Figure 1.29 shows a comparative chromatogram of blood after oxygenation and before oxygenation. In this tracing, one notes the increased oxygen content but loss of carbon dioxide after oxygenation. A measured amount of air (6 μl) was added to both samples to illustrate the effect of air contamination on the chromatogram.

Regarding the oxygen content of blood, specific studies have been conducted by Chambliss and Nouse [61], Lukas and Ayres [57], Russ et al. [58], and Ramsey [56].

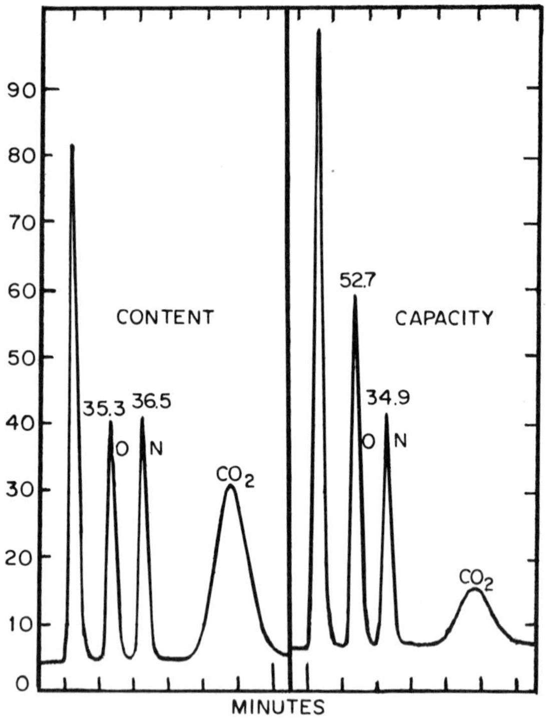

Figure 1.29. Tracing of same blood for oxygen before and after blood has been saturated with oxygen. From Natelson and Stellate [55], courtesy of Analytical Chemistry.

In 1959, Ramsey called attention to the applicability of gas chromatography to the analysis of gases contained in biological fluids. Using instrumentation and apparatus consisting of a gas chromatograph (Beckman GC-2A) and a Van Slyke volumetric extraction chamber, the waste arm of which was connected to the carrier-gas stream of the gas chromatograph proximal to the chromatographic column, he demonstrated that the oxygen extracted from water and plasma that has previously been equilibrated with room air could be measured with great accuracy and ease. The oxygen tension of water and plasma as determined by gas chromatography is shown in Table 1.7. As noted by Ramsey, such determinations previously required exceptional technical skill and even then were less accurate and more time consuming than determinations made according to the method described.

TABLE 1.7

Oxygen Tension of Water and Plasma as Determined by Gas Chromatography[a,b]

Type	Sample No.	Sample Size (ml)	Peak height (mm)	O_2 (ml)	P_{O_2} (mm Hg)
Room air	15	0.03	122 ± 0.17	0.0059	140
Water	35	1.0	124 ± 0.23		148 ± 0.28
Plasma	27	1.0	92 ± 1.04	0.0044	148 ± 1.68
Plasma (duplicate analysis)	13	1.0	92 ± 0.70	0.0044	148 ± 1.13

[a]From Ramsey [56], courtesy of Science.

[b]Oxygen tension of water and plasma as determined by gas chromatography. Column, 4-ft. 5A molecular sieve; current, 360 ma; flow pressure, 20 lb/in.^{2}; carrier gas, helium; column temperature, 100°C for water and 70°C for air and plasma. Standard deviations are given in columns 4 and 6.

Lukas and Ayres [57] explored the possibility of adapting the method proposed by Ramsey specifically to the determination of oxygen content of whole blood. Using a modified technique that nevertheless employed a borax-saponin-ferricyanide solution [72] for lysing the red cells, inactivating the hemoglobin, and preventing the release of CO_2 of the blood sample, and caprylic alcohol as an antifoam, agreement between chromatographic and Van Slyke values for oxygen contents was excellent ($r = 0.999$). Since an accuracy of approximately 1.2×10^{-5} ml can be achieved according to Ramsey [56] by his determination of the O_2 dissolved in 1 ml of water exposed to room air at 37°C (0.0044 ml), the use of quantities of blood for O_2 analysis much smaller than was used for their study was predicted by Lukas and Ayres.

In 1964, Russ, Carr, and Taylor [58] described a method of calibrating the chromatographic system for measuring blood oxygen content. Using a special reaction chamber and automatic chromatographic system of Fisher design, 10 ml of packed red blood cells were diluted with 10 ml of serum and saturated with oxygen. One-milliliter, 0.5-ml, and 0.25-ml samples of the diluted packed cells were analyzed for O_2 by the Van Slyke method. Whole blood was transferred to both the Van Slyke and the GC reaction chambers using 1.0- and 0.1-ml Hamilton syringes, respectively. By plotting the GC peak height of the chromatogram of the standard blood samples for oxygen versus volume percent of oxygen obtained for similar standardized samples by the Van Slyke technique, a linear relationship was found to exist. Examining 44 samples in duplicate by both methods, they found that the range of difference between the Van Slyke and the chromatograph was 0.00 to 0.50 vol. %, whereas the mean difference between the oxygen content of the unknown samples determined by both techniques was 0.20 (±0.02 S.E.M.) vol. %. The correlation coefficient calculated from this data was 0.97, indicating a high degree of relationship between oxygen values obtained by these two methods.

As reported, some gases occur in blood in bound form and must be liberated chemically. For example, CO_2 exists as bicarbonate and must be liberated with an acid, whereas CO and O_2 (both combined with hemoglobin) are released only after the blood specimen has been treated with a hemolyzing agent such as saponin and potassium ferricyanide together with an antifoaming agent. In the method described by Dominguez et al. [64], by which CO is released from carboxyhemoglobin by treatment with potassium ferricyanide reagent, the gas is extracted from the liquid phase with helium contained in a hypodermic syringe. For the determination of CO contained in the helium, an aliquot of the gas in the syringe is injected into the gas chromatograph where the CO is separated on a 2-m column packed with type 5A molecular sieve maintained at 75°C. With a helium flow rate of 135 ml/min, CO is eluted in about 5 min after nitrogen as a well-resolved peak. In their procedure, the parameters selected are such that methane, ethane, acetylene, hydrogen sulfide, carbon dioxide, hydrogen, and argon do not interfere. To compute the percent carboxyhemoglobin in the sample, analyses are made on the sample as received and on an equivalent sample saturated with carbon monoxide (carbon monoxide capacity). The percent carboxyhemoglobin in the sample is computed by comparing the peak height obtained on the unsaturated sample (carbon monoxide content) to the peak height of the saturated sample (carbon monoxide capacity) [73]. Measurable peaks were found with gas samples containing 5×10^{-4} ml of CO in 25 ml of helium.

The percent carboxyhemoglobin (CBH) is given by the expression

$$\text{CBH}\ \% = \frac{A \times V_S \times 100}{A_S \times V} \tag{1.3}$$

where

- A = peak height measurement of the unsaturated sample (carbon monoxide content)
- A_S = peak height measurement of the saturated specimen (carbon monoxide capacity)
- V = volume of unsaturated specimen
- V_S = volume of saturated specimen

TABLE 1.8

Ionization Potentials of Gases and Varied Organic Materials[a]

Material	Ionization potential
Helium	24.6
Argon	15.7
Hydrogen	15.6
Nitrogen	15.5
Carbon dioxide	14.4
Methane	14.4
Carbon monoxide	14.1
Ethane	12.8
Water	12.6
Oxygen	12.5
Acetylene	12.3
Cyclopropane	11.8
Ethylene	11.5
Argon (metastable)	11.5
Paraffins (C_3-C_{10})	10.5-11.2
Cycloparaffins (C_4-C_6)	10.5-11.0
Alcohols (C_1-C_3)	10.1-11.0
Aldehydes	10.0-11.0
Ketones	9.7-9.9
Olefins (C_3-C_6)	8.4-9.8
Aromatics	8.3-9.5

[a]From Guild et al. [60].

With this procedure, excellent agreement was obtained between the GC method and the Roughton-Scholander microgasometric method when blood samples containing various amounts of CO were analyzed.

The above method was later modified to permit the liberation of the gases with either the Van Slyke or Natelson microgasometer with a gas-sampling chamber attached to the GC unit by means of a special four-way stopcock [73]. The Van Slyke can be used with samples as large as 20 ml, whereas the Natelson can analyze samples as small as 0.01 ml.

Although all of the studies discussed utilized thermal conductivity detectors, it should be pointed out that ionization detectors are suitable for the detection of inert gases as shown in Figure 1.2. With helium as carrier gas, a direct ionization of the sample occurs with electron energies below the ionization potential of helium. No sample component has been found which could not be handled by the use of helium as carrier gas. As noted by Guild, Lloyd, and Aul [60], all of the materials listed in Table 1.8 can be readily determined. With ionization detectors, several orders of magnitude in sensitivity (about 1000 times) greater than those obtained with thermal conductivity-sensing devices can be achieved. This type of detection system would be most advantageous if one is specifically concerned with analyzing gases or volatiles at low concentration levels. It would also be preferred to the flame ionization detector, which yields no response to inert gases.

III. VOLATILE, LOW-MOLECULAR-WEIGHT ANESTHETICS

In addition to respiratory and blood gases, gas chromatography has been applied with great success to various problems associated with anesthesiology [3-5,8-12,38,74-126]. During anesthesia and in respiratory research, there is a need for analyzing mixtures of organic vapors on a routine basis; some mixtures at times contain components such as cyclopropane, chloroform, diethyl ether, divinyl ether, halothane, nitrous oxide, trichloroethylene, trifluoroethylvinyl ether, ethylene, ethyl chloride, methoxyflurane, 1,1,1-trichloroethane, tetrachloroethylene, halopropane, etc. In recent years, GC studies have been extended to areas concerned with the uptake, distribution, elimination, and administration of inhalation anesthetics, their partition coefficients in blood, tissues and lipids [10], the effects of pulmonary "first pass" of volatile or gaseous compounds administered intravenously on the calculation of apparent volume of distribution, pulmonary clearance, and determination of bioavailability from inhalation [11], as well as new, simple ways of studying protein binding of volatile and gaseous compounds [12].

Adlard and Hill [5] developed various chromatographic methods for the analysis of anesthetic mixtures. In respired mixtures containing CO_2, N_2O, and O_2, they found that a column 24 ft long x 1/4 in. i.d. packed with 28.6% propylene carbonate on 52-60 B.S. mesh Sil-O-Cel firebrick could adequately separate this mixture at room temperature in 25 min. However, a more rapid analysis can be achieved on a 20-ft x 1/4-in. column filled with 20% dimethyl sulfoxide on the same solid substrate: a mixture of oxygen, nitrous oxide, and carbon dioxide resolved in a total analysis time of 1 min. If halothane and diethyl ether are also present, the dimethyl sulfoxide column is used in parallel with a 2-ft x 1/4-in. column containing 15% dinonyl phthalate on firebrick. Using automatic valves to inject different-sized samples simultaneously into both columns, the dinonyl phthalate (maintained at 75°C) permits rapid elution of the light components but retains ether and fluothane (halothane) long enough for complete separation. With this dual-column arrangement, the effluents from the two columns pass through opposite channels of a thermistor thermal conductivity detector. However, the bridge circuit is modified so that a negative signal actuates a reversing switch so that only positive peaks are observed on the chart paper, as shown in Figure 1.30 for the separation of a respired anesthetic mixture, which can be performed in 45 min.

In work on the uptake and elimination of volatile anesthetics by the body, Butler and Hill [4] developed a method by which the anesthetic is extracted from blood (or other tissue) into n-heptane; the n-heptane extract is subsequently analyzed by gas chromatography. They found that it is possible to estimate halothane in n-heptane at concentrations less than 10 ppm. The separation is carried out with a 12-in. x 1/4-in.-i.d. copper column packed with 10% (w/w) of silicone fluid M.S. 550 impregnated on 80-100 B.S. mesh Celite. Using a hydrogen carrier gas flow rate of nearly 45 ml/min, the column temperature is held at 25 to 30°C for diethyl ether and halothane; for substances with higher boiling points (for example, chloroform, trichloroethylene, and methoxyflurane), a temperature of 100 to 120°C is suggested.

Using this procedure with a modified flame ionization detector, which appeared to be more sensitive to halogenated compounds than the usual flame detector [127,128], a halothane calibration curve had a coefficient of variation of 0.68%, with halothane providing a measurable peak at a concentration of about 10 ppm. At 30°C with a carrier gas flow of 45 ml/min, diethyl ether, halothane, n-heptane, and trichloroethylene in an extract had retention times of approximately 2.5, 4.5, 11.0, and 14.5 min, respectively.

Hall, Norris, and Downs [74], concerned with the physical chemistry of halothane-ether mixtures, showed that the azeotrope could be completely and easily separated into its component parts by GC on either a 12-ft benzyl

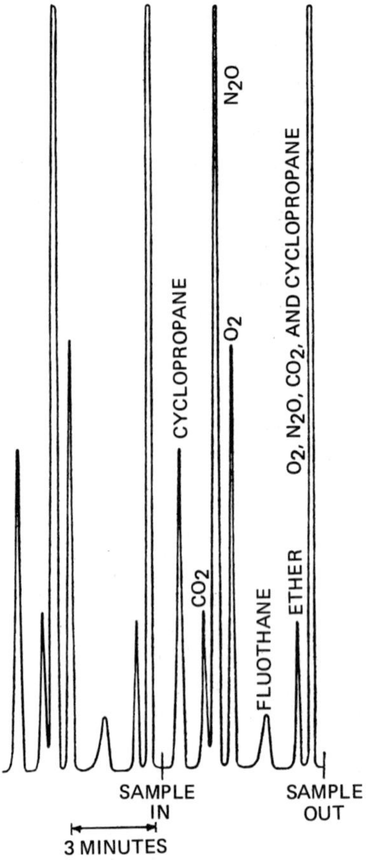

Figure 1.30. Chromatogram of a respired anesthetic mixture. Column 1, 20 ft x 1/4 in., 20% dimethyl sulfoxide on 52-60 mesh Sil-O-Cel at 20°C; inlet pressure 280 mm Hg. Column 2, 2 ft x 1/4 in., 15% dinonyl phthalate on 52-60 mesh Sil-O-Cel at 75°C; inlet pressure 40 mm Hg. Flow rate, 30 ml/min. H_2 in both columns. From Adlard and Hill [5], courtesy of Nature.

ether or a 2-m diisodecylphthalate column at 40°C with helium as carrier gas. In their preliminary study, two gas chromatographs (Beckman GC-2 and PE 154-C) were used, each equipped with thermal conductivity cells. Other investigators have also studied these and other compounds, singly or in mixtures, in blood, gas, or tissue.

Noehren and Cudmore [3], Baker and King [76], and Northington and Owens [80] determined blood ether levels by gas chromatography. In their GC method, Noehren and Cudmore extracted the gas in the tonometer of a Van Slyke apparatus. The diethyl ether content in blood was obtained using a PE 154-C Vapor Fractometer equipped with a gas-sampling valve accessory and a 2-m SS column packed with a diatomaceous earth support coated with tetraethyleneglycol dimethyl ether. With helium as carrier gas and the column maintained at 60°C, results could be obtained within 15 min; the accuracy within the concentration range of 25 to 175 mg% was ±3.05 mg and ±4.57 mg at the 95 and 99% confidence limits, respectively.

Baker and King [96] chose a solvent-extraction approach in preference to manometric extraction or direct injection of the blood sample in order to minimize inaccuracy. Carbon tetrachloride was used in a 1:1 ratio with blood, which was drawn and handled anaerobically until extraction with the solvent. To 1 ml of CCl_4 contained in a 3 to 5 ml stoppered vial or Vacutainer, an equal amount of blood was transferred directly into the vial by attaching a needle to the syringe and puncturing the stopper. After several minutes of shaking and centrifuging at high speed, a 10-μl sample (removed anaerobically using a microliter syringe) was injected into a Beckman GC-2 gas chromatograph with a Carbowax 400 column material used to achieve the separation of ether and CCl_4. When blood was extracted with CCl_4, it was possible to recover 95% of the ether, with a standard deviation of 1.5%. At the 99% confidence limit, the overall accuracy of the method was 3.9%.

On the other hand, Northington and Owens [80] used a Beckman GC-2 chromatograph equipped with a standard partitioning column, 6-ft x 1/4-in.-i.d. SS tubing packed with 42-60 mesh C-22 firebrick, coated with 50% (w/w) Dow-Corning silicone fluid (type 550). At a column temperature of 100°C and a helium flow rate of 101 ml/min, the ether is eluted in less than 1.5 min and all other sample components are eluted in 10 min. However, in this study, the ether is extracted from blood into cyclohexane solvent. This is accomplished by a modification of the extraction and "salting-out" procedure used by Cadman and Johns [129] for the determination of blood ethanol. Following detailed, precise calibration and extraction procedures, a statistical analysis of the results obtained in the preparation of 93 standard samples yielded a standard deviation from the curves ranging from 0.10 to 3.81 mg%. The average standard deviation is 1.93 mg%, determining an approximate error at the 95% confidence level of ±3.86 mg%.

Cheneau and Guichon [78] conducted studies in 1963 to find a suitable stationary phase which would allow analysis of an anesthetic mixture consisting of ethyl chloride, ether, and chloroform to be performed in approximately 5 sec. Squalane proved to be the most suitable liquid stationary phase for this rapid-analysis investigation. Using a 1-m by 4-mm column packed with 1% squalane impregnated on 110-120 mesh Anakrom

ABS (acid- and base-washed Chromosorb W, treated with hexamethyldisilazane) and a flow rate of 308 ml/min (column maintained at 5°C), an excellent separation was achieved in 20 sec. By increasing the flow rate to 1240 ml/min, the analysis time was reduced to 5 sec. A satisfactory separation was still obtained even at this great elution speed. However, to achieve these results, a special catharometer detector of weak time constant was used (approximately 0.01 sec) in addition to S.E.F.R.A.M. registering equipment. Samples were injected in vapor form by a special automatic sampler.

In their discussion of an analytical GC method for anesthesiology research, Summers and Adriani [75] review sampling methods for blood gas analysis and make special recommendations for the analysis of specific gases: (1) For low-boiling gases, such as N_2O, ethylene, and cyclopropane, use silica gel, methylhexyl sebacate, or dodecyl phthalate columns at room temperature to achieve elution times under 3 min, (2) for the separation of divinyl ether and diethyl ether with diethylhexyl sebacate or dodecyl phthalate, use columns operated at temperatures that approach the boiling points of the ethers in order to obtain elution times of about 5 to 10 min; and (3) for chloroform, halothane, and trichloroethylene, use silicone oil as the liquid stationary phase at 100 to 110°C to obtain elution times between 7 and 10 min.

Zauder and Orkin [77,79] found in their quantitative study of anesthetic agents that, for routine use, a 2-m silicone oil column (DC 200) operated at 75°C with a helium flow of 90 ml/min was preferred. For common anesthetics such as chloroform, cyclopropane, diethyl ether, divinyl ether, halothane, nitrous oxide, trichloroethylene, and trifluoroethyl vinyl ether, the relationships of peak height and area to concentration were evaluated. In this regard, they noted that approximation of the peak area by multiplying the peak height by its width at one-half the peak height gave less reliable results than electronic integration of peak area or measurement of the peak height.

Using an F&M flame ionization detector and electrometer and a special thermostated, solenoid-timer injection port system installed between the carrier-gas flow regulator and the regular injection port, Lowe [81] injected 1-μl samples or aqueous standards when the carrier gas flow bypassed the injection port. After injection of the sample, the timer was activated to allow 15 sec for volatilization followed by 15 sec of purging and automatic resetting for a subsequent sample. With this arrangement, the injection port was kept between 56 and 60°C when analyzing 1-μl samples but was maintained between 100 and 120°C when weighed specimens were to be determined for their anesthetic content. Tissue and blood (heparinized or clotted) were extruded from syringes into weighed glass capillaries using a Becton adapter. If the analysis could not be performed within 2 to 3 hr, the specimen was extruded from the syringe into flux-free solder tubing and both ends crushed with pliers. To retrieve the tissue for analysis, one end was opened by scratching deeply with a file and flexing until a clean break was obtained. Following the transfer of the sample into a weighed glass capillary tube, the glass capillary with tissue was reweighed and analyzed as described above. Columns packed with glass wool or Chromosorb P (1-ft x 1/4-in. Polyflo tubing) and saturated with water were employed to stabilize the vapor pressure of water entering the flame, thus eliminating the fluctuations produced by vaporization of injected samples.

The columns used to achieve separation of specific mixtures were as follows: (1) a 1-ft column of wet Chromosorb P was suitable for the separation of cyclopropane-ether or cyclopropane-halothane mixtures without further modification; (2) 1-ft columns of dry Chromosorb P and wet glass wool in series resolved mixtures of cyclopropane, halothane, and ether; (3) a 2-ft column of Tide and a 1-ft column of wet glass wool in series separated cyclopropane, halothane, ether, chloroform, and methoxyflurane (penthrane); (4) a 2-ft column of silicone rubber in series with a 1-ft column of wet glass wool could separate cyclopropane from ether or halothane but failed when applied to halothane-ether mixtures; and (5) a 9-ft Tide column at 100°C was suitable for the complete separation and analysis of a mixture of seven commonly used anesthetic agents (C_3H_6, fluoromar, ether, halothane, $CHCl_3$, trichloroethylene, and methoxyflurane).

Using such a technique, Lowe [81] summarized its advantages in the following manner:

> The ability to inject gases, blood, and tissue directly into the injection port eliminates extraction or equilibration procedures. In addition, the injection technique is amenable to anaerobic handling of specimens which greatly minimizes losses of highly volatile materials. Samples may be stored for two to three weeks in solder tubing without loss of anesthetic vapors. Where more than one organic vapor is present, these may be easily separated by appropriate columns which only prolong the analysis time by a few minutes. The conditions for operating the flame detector are the same for all anesthetics so that determinations for different agents can be run in sequence. The small size of the samples required for analysis permits the collection of finger or ear lobe blood for repeated analysis. As many as 100 to 200 one microliter blood samples may be analyzed without cleaning the injection port, although the dried crust may slow down the distillation slightly. This can be avoided partially by varying the depth of injection. Eight to 10 tissue analyses on 25 mg specimens in glass capillaries can be performed before cleaning of the injection port is required. The injection port is cleaned within one minute by flushing with water and blowing with air.

Reporting on a similar study as that undertaken by Lowe [81], Lowe and Beckham [83] prepared volumetric standards by adding 5 μl of liquid anesthetic to 20 ml of water contained in a capped syringe. The specific gravities listed in Table 1.9 were used to calculate the concentration in milligrams per 100 ml. For all anesthetics, the response of the flame ionization detector was linear over a 100,000-fold range in concentration (8 x 10^5 mg/100 ml to 1 mg/100 ml of halothane). Data were presented showing the rise in blood halothane concentrations during inductions with halothane and oxygen. A sharp decline in blood concentrations occurred following the administration of succinyl choline, hyperventilation with oxygen, and intubation of the tracea. Additionally, the hypertensive action of halothane and its relationship to blood concentrations was discussed. If the individual patient's blood concentrations are plotted against the observed percent decrease in blood pressure (0% taken as the patient's normal preoperative blood pressure), the percent decrease in blood pressure from normal for each patient was proportional to the blood halothane concentration. The similarity of the slopes of the curves for all patients indicated a 10% decrease in blood pressure for each 2.5 mg/100 ml increase in blood concentrations above 5 to 8 mg/100 ml.

The approach taken by Gadsen and McCord [82] was essentially a preparative technique for quantitation of volatile materials (fluothane, fluoromar, alcohols) from blood by gas chromatography. The distillation technique involves precipitation of proteins (accomplished by sodium tungstate-sodium sulfate and sulfuric acid-sodium sulfate reagents), addition of a water-immiscible solvent, subsequent distillation, followed by dehydration (anhydrous potassium carbonate) of the distillate. The immiscible solvent is selected on the basis of several criteria: (1) relatively low boiling point, (2) its ability to "carry over" quantitatively the material being analyzed while being distilled, (3) its ability to be separated chromatographically from the material(s) under investigation. As examples, Gadsen and McCord selected toluene (for fluothane and fluoromar) and n-butyl alcohol (for the alcohols). The isolation, detection, and quantitative measurement could be accomplished in approximately 30 min for a single sample.

Using toluene as the distillation "carrier," the fluothane-toluene and fluoromar-toluene distillates were chromatographed with a Beckman GC-2 instrument equipped with a 24-ft column filled with 10% (w/w) di-2-ethylhexylsebacate on 20-60 mesh firebrick. For a 25-μl injection of distillate, with the column operated at 130°C, the retention times for air, fluoromar, fluothane, and toluene were approximately 0.5, 1.25, 2.75, and 7.5 min, respectively. Results of some tests of reproducibility and recovery are presented in Tables 1.10 and 1.11, respectively.

TABLE 1.9

Physicochemical Properties of Anesthetics[a]

Agent	Formula	Molecular weight	Vapor STP (mg/100 ml)	Vapor pressure, 25°C (mm Hg)	Ostwald partition coefficient	Anesthetic blood concentration (mg/100 ml)
Cyclopropane	C_3H_6	42.08	1.88	90 psi	0.415	8–18
Ether	$(C_2H_5)_2O$	74.13	3.31	540	14.4	50–100
Methoxyflurane	$C_2F_2Cl_2HOCH_3$	165.0	7.35	33	8.4	10–20
Halopropane	$C_3F_4ClH_3$	150.0	6.70	95	5.4	10–20
Halothane	C_2F_3HBrCl	197.4	8.78	300	2.36	10–20
Chloroform	$CHCl_3$	119.4	5.34	200	7.2	10–20
Fluoromar	$C_2F_3H_2OC_2H_4$	131.4	5.86	380	1.56	10–20

[a]From Lowe and Beckham [83], courtesy of Plenum Press.

TABLE 1.10

Reproducibility Experiments[a]

Sample no.	Peak ht. (in.)	Conc. (mg%)	Ave. conc. (mg%)	Percent ave. conc.
Fluothane--Blood Samples (4 cm^3 distilled--triplicates)				
A	40.0	20.6	20.2	102.0
	38.9	20.0		99.0
	38.5	19.9		98.5
B	27.1	14.0	14.5	96.5
	29.3	15.1		104.1
	28.0	14.4		99.3
C	77.2	39.8	40.2	98.9
	79.1	40.8		101.5
	77.8	40.1		99.7
Fluoromar--Blood Samples (4 cm^3 distilled--triplicates)				
A	89.2	74.7	73.8	101.2
	88.6	74.2		100.5
	86.7	72.6		98.4
B	52.1	43.7	44.4	98.3
	54.2	45.4		102.2
	52.8	44.2		99.5
C	20.8	17.4	18.3	95.0
	22.3	18.7		102.2
	22.6	18.9		103.3

[a]From Gadsen and McCord [82], courtesy of Journal of Chromatographic Science.

Many of the halogenated anesthetics have been extensively investigated by gas chromatography. For example, the purification of halothane-1-^{14}C, halothane tensions in blood, its impurities, its determination in tissue, and the toxicity of dichlorohexafluorobutene (a trace component of halothane) have all been adequately handled by chromatographic techniques.

Halothane (2-bromo-2-chloro-1,1,1-trifluoroethane) has been determined in a variety of mixtures using many different column systems: some of these include (1) that used by Adlard and Hill [5] previously discussed in this chapter, (2) those reported by Lowe [81,108] as previously cited, (3) a 6-ft x 0.062-in. i.d. column filled with approximately 23% (w/w) MS 550 silicone fluid coated on 60-80 mesh Chromosorb P at 88°C [111], (4) a 6-ft x 3/16-in. molecular sieve column in parallel with a 12-ft x

TABLE 1.11

Recovery Experiments[a]

Sample no.	Conc. (μg)	Added (μg)	Total (μg)	Found (μg)	Recovery (%)
Fluothane[b]					
A	32.6	40.0	72.6	70.9	97.7
				74.1	102.1
B	57.6	40.0	97.6	101.2	103.6
				97.3	99.7
C	63.6	80.0	143.6	140.2	97.6
				146.2	101.8
Fluoromar[b]					
A	14.6	21.3	35.9	36.6	101.9
				36.8	102.5
B	29.9	35.0	64.9	65.8	101.3
				61.8	95.2
C	56.2	67.6	123.8	127.1	102.7
				123.3	99.6

[a]From Gadsen and McCord [82], courtesy of Journal of Chromatographic Science.

[b]The concentrations noted are based on the absolute micrograms of material per 25 μl injected onto the chromatograph column. The blood and standards were then mixed and redistilled and this distillate injected. These are then compared to the amount of substance (fluothane or fluoromar) expected.

3/16-in. column packed with 10% (w/w) Amine 220 and 5% Carbowax 400 (w/w) on 30-60 mesh Teflon-6 at 64.5°C [91], (5) a 60-cm x 3/16-in. column containing 20% (w/w) dioctyl sebacate on firebrick in parallel with a 120-cm x 3/16-in. column filled with 10% dioctyl sebacate on silica gel, both columns in turn placed in series with a 183-cm x 3/16-in. 5A molecular sieve column [110], (6) a 6-ft x 1/4-in. column packed with Tide detergent maintained at 90°C [90], and (7) a 50-cm x 4-mm-i.d. column packed with 15% (w/w) no. 10 Kel-F oil coated on 0.16- to 0.20-mm Chromaton NAW in series with a 270-cm-long column filled with 100-120 mesh Porapak (room temperature) [38].

Van Dyke, Chenoweth, and Larson [106] purified crude halothane-1-^{14}C by gas chromatography using a 5/8-in. x 10-ft bis-(2-ethylhexyl) sebacate on Chromosorb column at a temperature of 100°C. The 99.9% pure material was then used to examine the possibility that halothane is converted or metabolized to CO_2. This was undertaken because of evidence that had been accumulated in their laboratory that volatile anesthetics are not biologically inert, these authors having measured the extent of CO_2 production from ^{14}C-diethyl ether, ^{14}C-chloroform, and ^{14}C-methoxyfluorane. Their findings suggested that the extent of the number one carbon of halothane to CO_2 was very limited since only 0.84% of the injected dose appeared as CO_2 in 24 hr.

Purchase [8] estimated the tension of halothane in blood since this is directly related with the concentration of anesthetic found in the brain. Furthermore, at a given tension, the concentration in the blood is proportional to the blood/gas partition coefficient.

Using a modified version of a method described by Curry et al. [9], Purchase separated halothane with a 2-ft x 1/4-in. column filled with 20% (w/w) SE-30 silicone rubber coated on Chromosorb. With the flame ionization detector, the temperatures of the injection port and column were 125 and 100°C, respectively, whereas the argon carrier-gas flow rate was set at 45 ml/min. With these operational parameters, halothane was eluted in 22 sec.

Because of reports [130] of hepatic necrosis following the administration of halothane, several studies were initiated to determine the possible presence of other toxic halogenated compounds [84-89,113,114,116,118, 121].

Cohen et al. [84] noted that gas chromatographic analysis of halothane (FID detector; diisodecyl phthalate column; 70°C; helium flow, 50 ml/min) showed the presence of a compound with a retention time of 35 sec, a retention time much shorter than that for halothane (75 sec) under identical conditions. The average concentration of this impurity in freshly opened bottles was approximately 0.01%. By mass spectrometry, the compound was identified as 2,3-dichloro-1,1,1,4,4,4-hexafluorobutene-2, existing as the cis or trans isomer. They proposed the following chemical mechanism for its formation from halothane:

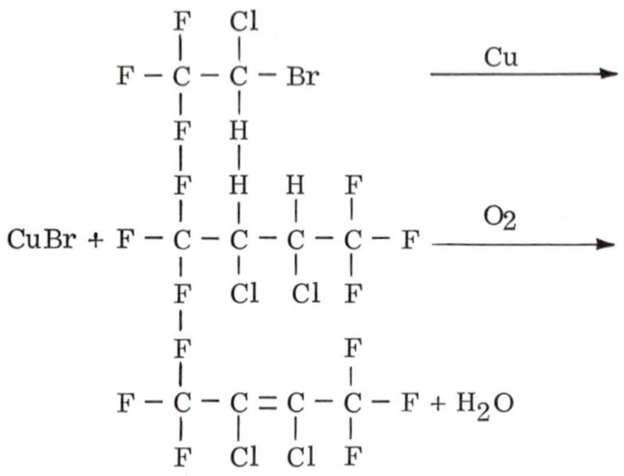

(1.4)

Lu et al. [131] noted that this compound, which is structurally related to fluorocarbons of high toxicity, produced convulsions at the point of anesthesia in two of four rats studied, whereas Clayton [132] has shown that the presence of a double bond in the polyfluoroalkenes is associated with an increase in chemical activity and toxicity over that shown by the alkanes.

To test the hypothesis proposed that halothane in the presence of copper and oxygen forms butene, Butler and Linde [87] examined their distillation residue, which contained the less volatile residue from 1 liter of old halothane removed from five copper kettles, examined 28 samples of halothane removed from 14 copper kettles, shook halothane, copper compounds, and oxygen for 6 to 8 hr with and without exposure to ultraviolet light, and evaporated the halothane in a stream of oxygen with and without copper filings. All samples were chromatographed on each of three columns (60-cm SE-30 column, 2-m 10% DC 200 silicone oil, 2-m Tide; temperatures from 80 to 120°C). All samples showed the presence of butene as well as three to five trace contaminants at concentrations ranging from 0.01 to 0.05%. From this study, they concluded that hexafluorodichlorobutene is not formed by a reaction involving copper and halothane and that the moderate increase in butene is due to the preferential evaporation of the more volatile halothane.

An examination of discolored and freshly opened halothane samples indicated the presence of five trace impurities as well as halothane and thymol (thymol added as a stabilizer at a concentration of nearly 0.01%). Using their chromatographic columns and conditions, a butene (reported to contain equal proportions of the cis and trans isomers) was analyzed and found to contain six major components, two of which represented the cis and trans isomers (47 and 49%, respectively). On each column, one or both of these isomeric butenes had peaks with retention times identical to those tentatively identified as butene in their previous halothane chromatograms on these columns. Their enrichment theory was also supported by Phillips [88] in his investigation of stored specimens as compared to fresh commercial samples. This conclusion was based on results obtained from a modified PE 154-C gas chromatograph equipped with a thermistor thermal conductivity (TC) and a hydrogen flame ionization (FID) detector placed in series and a 20-ft x 1/4-in. copper column packed with 20% (w/w) DC 550 silicone oil on 60-80 mesh Chromosorb P.

A typical chromatogram of the impurities noted by Cohen and co-workers [85, 89] from the injection of stock halothane is illustrated in Figure 1.31. The chromatogram shown in Figure 1.32 is the result of a new gas chromatographic technique proposed by Cohen and Brewer [85] for the analysis of anesthetic gases in tissue. Figure 1.32 presents a typical chromatogram obtained from the direct injection of 11 mg of brain tissue sample from a monkey anesthetized for 1 hr with a mixture consisting of 1% halothane and 0.1% dichlorohexafluorobutene. Their technique has been used satisfactorily for the direct injection of brain, kidney, lung, fat, liver, muscle, pancreas, thyroid, heart, and adrenal tissues.

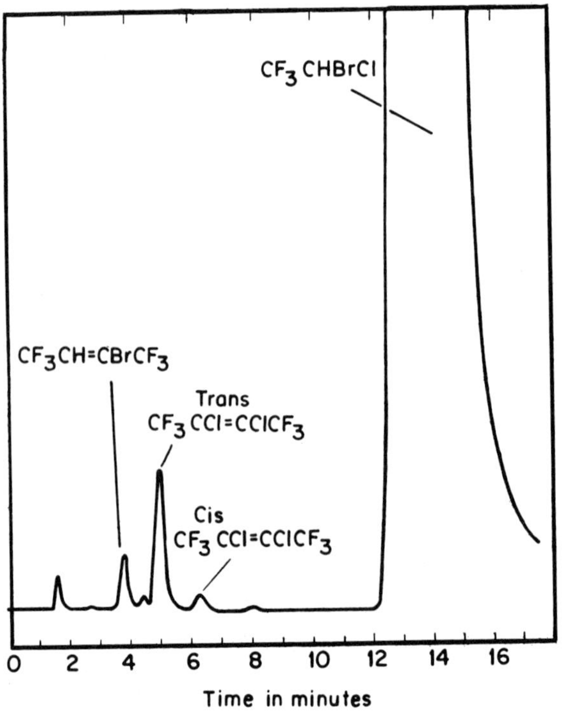

Figure 1.31. Chromatogram obtained from the injection of 1 μl of stock halothane (Ayerst Laboratories, lot D7892HF). Note separation of the three "butene" impurities. From Cohen et al. [89], courtesy of Anesthesiology.

In 1967, Chapman et al. [116] studied the chemical aspects and impurities in halothane produced by high-temperature bromination of 2-chloro-1,1,1-trifluoroethane. Using other complementary analytical instrumentation for the identification of impurities, they found and identified 16 contaminants in crude halothane. Gas chromatographic methods were developed to determine quantitatively their concentrations in the final product at the parts-per-million level. For this problem, two chromatographic systems were devised. System A consisted of a 6-ft x 3/16-in.-i.d. column filled with 30% (w/w) Aroclor 1254 coated on 60-80 mesh Chromosorb P; column temperature, 60°C; nitrogen carrier-gas flow rate, 30 ml/min; flame ionization detector. System B consisted of a 6-ft x 3/16-in.-i.d. column packed with 30% (w/w) PEG 400 (polyethylene glycol) on 72-85 mesh Celite connected in series with a 3-ft x 3/16-in.-i.d. column containing 30% (w/w) dinonyl phthalate on 72-85 mesh Celite; column temperature, 60°C; nitrogen carrier-gas flow rate, 40 ml/min; flame ionization. The identification and

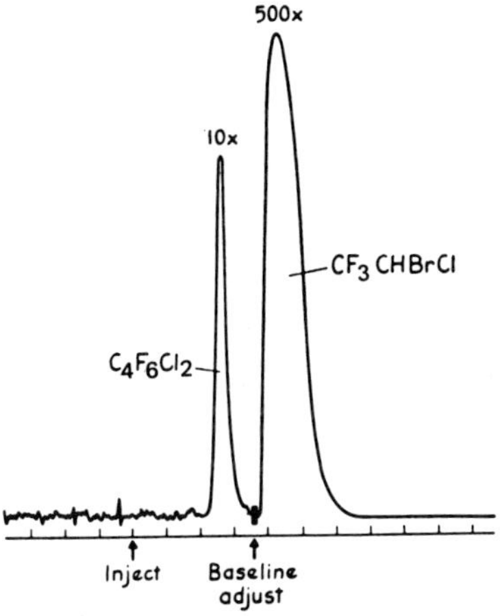

Figure 1.32. Chromatogram, brain tissue (monkey), 11 mg; column, D.C. 550; atten., 10X, 500X; carrier, argon, 5% methane; total flow, 300 ml/min (sensing column 30 ml/min); detector, electron capture; inj. temp., 125°C; col. temp., 72°C. From Cohen and Brewer [85], courtesy of Journal of Chromatographic Science.

retention times of the 16 impurities relative to 1,1,2-trichloro-1,2,2-trifluoroethane (identified as impurity no. 9) are listed in Table 1.12. Although both systems failed to separate compounds 5 and 6, only four were present in purified halothane as impurities that could be reasonably detected, these being compounds 2, 5, 9, and 14, for which average or typical concentrations of 1, 8, 12, and 1 ppm, respectively, were found.

The determination of halothane levels in expired air, body fluids, and tissue have been the concern of many investigators [4,8,81-83,85,90,91, 94,103-105,108,111,115,119,120,122-125].

In the analysis of halothane in gas (vapor) and blood, Rutledge et al. [90] facilitated the establishment of calibration curves using a set of equations for the calculation of the halothane vapor (gas) and its concentration in blood:

1. Vapor analysis calibration:

$$C_{HT} = \frac{(P_{HT})\ (100)}{(P_{At})} \tag{1.5}$$

$$P_{HT} = \frac{(n)\ (R)\ (T)}{(V)} \tag{1.6}$$

$$n = \frac{(\text{ml of HT})\ (\text{sp. gr. of HT})}{(\text{MW of HT})} \tag{1.7}$$

TABLE 1.12

Identity and Relative Retention Times of Some Possible Halothane Impurities[a]

Impurity number	Identity	Retention times relative to $CF_2ClCFCl_2$: Chlorinated diphenyl column[b]	Retention times relative to $CF_2ClCFCl_2$: Polyethylene glycol-dinonyl phthalate column[c]
1	trans-2-Chloro-1,1,1,4,4,4-hexafluoro-2-butene	0.12	0.38
2	2-Chloro-1,1,1-trifluoroethane	0.22	1.00
3	cis-2-Chloro-1,1,1,4,4,4-hexafluoro-2-butene	0.30	0.73
4	trans-2-Bromo-1,1,1,4,4,4-hexafluoro-2-butene	0.40	1.00
5	2,2-Dichloro-1,1,1-trifluoroethane	0.62	2.35
6	2-Bromo-1,1,1-trifluoroethane	0.62	2.35
7	trans-2,3-Dichloro-1,1,1,4,4,4-hexafluoro-2-butene	0.62	0.73
8	cis-2,3-Dichloro-1,1,1,4,4,4-hexafluoro-2-butene	0.78	0.85
9	1,1,2-Trichloro-1,2,2-trifluoroethane	1.00	1.00
10	Bromodichlorofluoromethane	3.48	ND
11	2,2-Dibromo-1,1,1-trifluoroethane	4.85	ND
12	Chloroform	6.43	10.9
13	2,2-Dibromo-2-chloro-1,1,1-trifluoroethane	8.73	ND
14	2-Bromo-2-chloro-1,1-difluoroethane	M	1.50
15	2-Bromo-2,2-dichloro-1,1,1-trifluoroethane	M	3.28
16	1,2-Dichloro-1,1-difluoroethane	M	4.22

[a]Adapted from Chapman et al. [116].
[b]M = peak masked by halothane.
[c]ND = not determined.

where

C_{HT} = gaseous concentration of halothane (HT), in vol. %
P_{HT} = partial pressure of halothane, in mm Hg
P_{At} = atmospheric pressure, in mm Hg
n = number of moles of halothane in the flask
R = ideal gas constant
T = absolute temperature
ml of HT = volume of liquid halothane injected in ml
sp. gr. of HT = specific gravity of halothane
MW of HT = molecular weight of halothane

2. Blood analysis calibration:

$$C_{HT} = \frac{(\mu l \text{ of HT})\ (\text{sp. gr. of HT})\ (100)}{(V_{Bl})} \tag{1.8}$$

where

C_{HT} = halothane concentration in the blood, in mg% (w/v)
μl of HT = volume of halothane added to the blood, in μl
V_{Bl} = volume of blood in the tube, in ml (calculated from weight/sp. gr.)

As a practical application of the method outlined by Rutledge et al., Figure 1.33 gives the results for an experiment where halothane concentrations in inspired and expired gas and in the blood from the venous reservoir of a dog heart-lung preparation were determined at intervals, and at two different settings of the cardiac output. They noted that the time course of change in halothane concentration corresponded very well with results expected from theoretical considerations of uptake and distribution of an inert gas with characteristics of halothane. Furthermore, a value of 3.55 for the blood/gas distribution coefficient compared favorably with that reported in the literature [133].

Theye [91] performed similar chromatographic analyses of expired air containing halothane with an accuracy acceptable for studies of respiratory and anesthetic gas exchange during halothane anesthesia, whereas Epstein et al. [92] showed the uptake of halothane (by gas chromatographic means) given in constant concentration (1%) to be more rapid in a mixture with 70% N_2O than in one with 10% N_2O. This finding demonstrated that the gas uptake is influenced by the constituents in the mixture of which it is a component. They noted that changes in other component concentrations sufficient to produce a significant concentration effect will also alter the uptake rate of companion gases given in constant concentration ("second gas effect"). They interpreted this acceleration of uptake to be a result of additional inspiratory inflow secondary to the absorption of N_2O at a higher concentration. In turn, this absorption produces the concentration effect on the rate of uptake of the N_2O itself as indicated by the data.

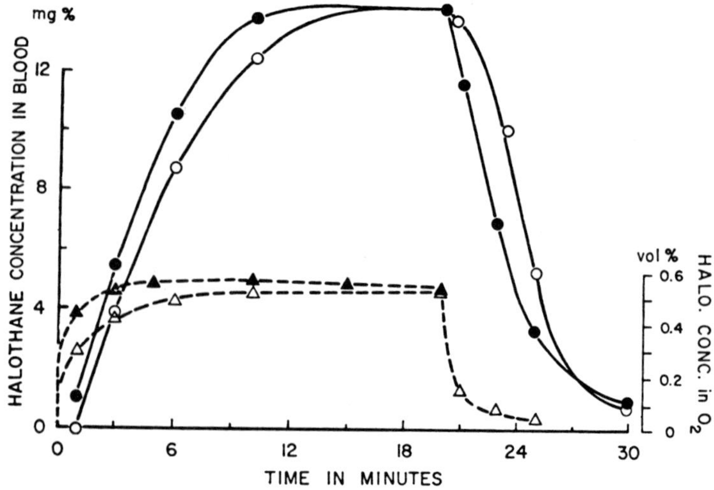

Figure 1.33. Time course of change of halothane concentration in inspired and expired gas and in blood of dog heart-lung preparation. Weight of dog, 11.5 kg. Estimated total blood volume, 1050 ml. Halothane administered at constant inspired concentration for 20 min at Fluotec setting of 0.5% by nonrebreathing system. Tidal volume = 350 ml at a rate of 20/min. Full circles: blood concentration at cardiac output of 730 ml/min. Open circles: blood concentration at cardiac output of 370 ml/min. Full triangles: inspired concentration. Open triangles: expired concentration. From Rutledge et al. [90], courtesy of Anesthesiology.

The applicability of the electron-capture or affinity detector to the analysis of halogenated hydrocarbons in expired air and blood has been investigated by Stewart et al. [93], Cohen and Brewer [85], Douglas et al. [119], Davies and Mathias [117], and Brachet-Liermain, Ferrus, and Caroff [94].

Stewart et al., concerned primarily with the detection of trichloroethylene, trichloroethane, and tetrachloroethylene, conducted a series of timed human exposures to known concentrations of these chlorinated hydrocarbons in an exposure chamber measuring 32 ft x 13 ft x 8 ft. This room had a continuous, positive air supply. The solvent was metered into the room and its atmospheric concentration determined simultaneously by three methods: (1) infrared analysis using a Perkin-Elmer model 12C IR spectrophotometer equipped with a 10-m path-length gas cell; (2) using a modified Davis halide meter; and (3) by adsorption of known volumes of air on silica gel, with subsequent combustion and analysis by Volhard titration. Gas chromatography was used to separate the components of the breath samples, an electron-capture detector determining the concentration of the chlorinated solvent by comparing its peak height with a standard curve. The Aerograph Hy-Fi A600B gas chromatograph was equipped with a 6-ft x 1/8-in.-o.d. SS column packed either with Apiezon L or Carbowax 20M on alkaline Chromosorb R. Having adjusted the carrier-gas flow rate and column temperature to obtain the best separation of trichloroethylene, 1,1,1-trichloroethane, and tetrachloroethylene, the limits of detection for trichloroethylene in air ranged from 2 to 10 ppb whereas, with a standard of 1.35 ppm trichloroethylene in air, the peak height was reproducible with a 2σ of less than one scale division in 60. The limits of detection for 1,1,1-trichloroethane, or tetrachloroethylene in air, ranged from 1 to 5 ppb. Following exposure, trichloroethylene was detectable in breath for 330 hr. On the other hand, the other two chlorinated compounds were detectable for 800 hr after exposure.

Brachet-Liermain et al. used electron-capture detection to measure halothane in blood. Using a benzene extraction of the halothane from blood, 1 μl of extract was injected into an Aerograph Moduline 1520 instrument with an ECD detection system (250 mCi tritium source); the separation was performed with a 15-in. x 1/8-in. SS column packed with Tide. With a carrier-gas flow rate of 25 ml/min and a column temperature of 75°C, levels of halothane in arterial, venous, and capillary blood were determined simultaneously during a study of saturation and desaturation of blood. The three blood samples were obtained simultaneously at the end of 30, 45, and 60 min. Table 1.13 lists the halothane levels determined in each type of blood.

TABLE 1.13

Halothane Blood Levels[a]

	Arterial		Capillary		Venous	
Time (min)	Halothane (mg/100 ml of blood)	Area (disk units)	Halothane (mg/100 ml of blood)	Area (disk units)	Halothane (mg/100 ml of blood)	Area (disk units)
30	1.60	388	1.03	250	0.66	165
45	1.60	378	1.33	310	0.78	185
60	1.61	370	1.33	315	0.87	200

[a]From Brachet-Liermain et al. [94], courtesy of Journal of Chromatographic Science.

Other halogenated organics have also been determined in air and body fluids: for example, halopropane (3-bromo-1,1,2,2-tetrafluoropropane) [96], methoxyflurane (2,2-dichloro-1,1-difluoroethyl methyl ether) [97], and fluorochlorocarbons (FCC 11, CCl_3F; FCC 12, CCl_2F_2; FCC 113, $FCl_2C\text{-}CClF_2$; FCC 114, $F_2ClC\text{-}CClF_2$) [98].

Jacobs [96] used two GC techniques: (1) direct injection of untreated whole blood onto a chromatographic column and (2) rapid, single-step extraction of the halopropane from blood into an equal volume of perchloroethylene with subsequent GC analysis of the extract. The first of these methods permitted the determination of as little as 5.0 mg of halopropane per 100 ml of blood within 90 sec (Fig. 1.34), whereas the second procedure required 30 sec for the analysis, with complete recovery of the halopropane over the range 2.5 to 200 mg/100 ml of blood (Figure 1.35).

For the direct injection approach, the column was composed of two parts: (1) 16 in. of phosphorus pentoxide powder (90% w/w) supported on 60-80 mesh firebrick (10% w/w) which had been pretreated with Desicote and Siliclad, and (2) 12 in. of 20% (w/w) DC 710 silicone oil supported on 60-80 mesh firebrick which had been pretreated with Desicote. Connected in series, the columns were maintained at 40°C with a carrier-gas flow rate adjusted to 40 ml/min. As shown in Figure 1.34, the halopropane was eluted in approximately 1.5 min.

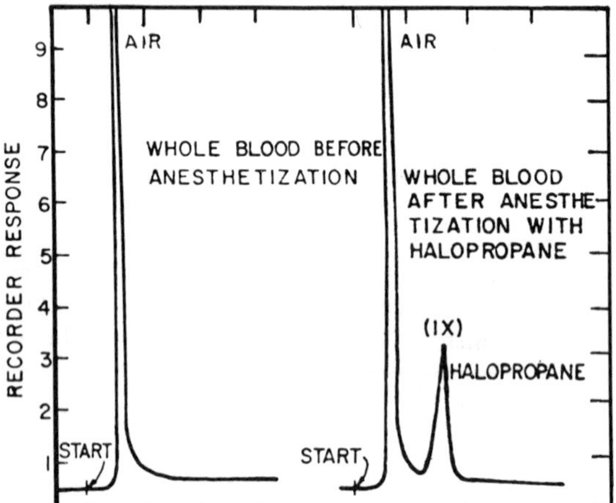

Figure 1.34. Sample chromatogram of blood before and after anesthetization with halopropane - direct injection procedure. From Jacobs [96], courtesy of Anesthesia and Analgesia.

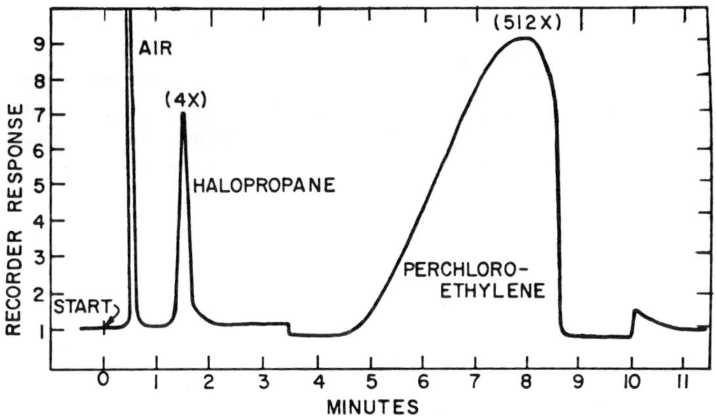

Figure 1.35. Sample chromatogram of halopropane in perchloroethylene--solvent extraction procedure. From Jacobs [96], courtesy of Anesthesia and Analgesia.

The solvent extraction method used was designed to determine halopropane in blood using the same silicone oil partition column (designated 2) found suitable for direct on-column injection of the sample. The water absorber column (1) is necessary only when the direct injection method is used. Although the precision of the direct injection method was acceptable (Table 1.14), the recovery was only about 60%. On the other hand, the halopropane was completely recovered by the single-step extraction of whole blood with an equal volume of perchloroethylene. Its precision is also given in Table 1.14.

Gadsen et al. [97] isolated methoxyflurane from 2- to 4-ml samples of blood by a distillation technique. Sample preparation consisted of adding isoamyl alcohol to the blood and then distilling the mixture through a Vigreaux column. Following the collection of the organic fraction of the distillate (iso-AmOH and methoxyflurane) and its dehydration over anhydrous K_2CO_3, a 20-μl aliquot was injected into a Beckman GC-2 gas chromatograph equipped with a 4-ft x 1/4-in. copper column packed with hexadecane coated on 30-50 mesh firebrick (Beckman no. 17296). With a helium flow rate established by a 40-psi inlet pressure and the column held at 70°C, the methoxyflurane and isoamyl alcohol were eluted in 3.25 and 6.30 min, respectively. With this method, analytical data obtained for dogs correlated well with values reported by Chenoweth et al. [134].

A gas chromatographic method for the separation and determination of fluorochlorocarbons (FCCs) was described by Rauwns et al. [98] based on the headspace technique used by Fink and Morikawa [125] for volatile anesthetics and by Van Stekelenburg and De Bruijn [135] for acetone in

TABLE 1.14

Determination of Halopropane in Blood[a]

A. Recovery of Halopropane in Blood by Direct Injection Gas Chromatography

Sample	Number of runs	Halopropane per 100 ml of blood (mg) Added	Found	Percent recovery
1	2	184.5	109.0	59.1
2	8	104.2	64.1	61.5
3	2	64.0	38.0	59.4
4	2	46.0	28.1	61.1
5	3	26.5	16.0	60.4
6	2	18.2	10.6	58.2

Weighted recovery average, 60.46%.
Standard deviation of means, ±1.3%.

Precision of Direct Injection Procedure for Determining Halopropane in Blood

Number of determinations	8
Halopropane found, mg/100 ml of blood	64.1
Standard deviation, (6), mg/100 ml of blood	±3.68
Ninety-five percent confidence limits for the mean of duplicate determinations, mg/100 ml.	±6.0

B. Determination of Halopropane in Blood by Solvent Extraction-Gas Chromatography

Sample	Number of runs	Halopropane per 100 ml (mg) Added to blood	Found in Perclene	Percent recovered
1	2	3.78	3.78	100.0
2	2	7.97	8.14	102.0
3	2	48.8	48.3	99.0
4	2	67.8	67.2	99.2
5	2	86.9	85.6	98.5
6	8	104.2	104.8	100.6
7	2	118.6	120.2	101.4
8	2	129.0	130.0	100.8
9	2	159.0	160.3	100.8

Weighted recovery average, 100.3%.
Standard deviation of means, ±1.1%.

TABLE 1.14 (continued)

Precision of Solvent Extraction Procedure	
Number of determinations	8
Halopropane found, mg/100 ml of blood	104.8
Standard deviation, mg/100 ml of blood	± 0.96
Ninety-five percent confidence limits for the mean of duplicate determinations, mg/100 ml	± 1.61

[a]From Jacobs [96], courtesy of Anesthesia and Analgesia.

blood. Dollery et al. [136] and Paterson et al. [137] determined FCCs in blood by direct on-column injection of blood samples, whereas Shargel and Koss [138] used a low-temperature procedure, which should prevent losses that normally occur when extraction is carried out at room temperature.

Using a Varian Aerograph 1440 equipped with a tritium electron-capture detector and a 150-cm x 0.3-cm-o.d. glass column filled with 100-120 mesh Porapak Q (column temperature held at 150°C), the separation of the four compounds is shown in Figure 1.36 (separation was accomplished with the nitrogen flow rate held at 25 ml/min). Table 1.15 contains, in addition to the retention times, their respective distribution coefficients in both blood and water and their detection limits. Determined in less than 10 min, the individual FCC components are eluted from the Porapak column according to their boiling points.

Although Lowe and Hagler [10] applied gas chromatography to the determination of the water-gas and blood-gas partition coefficients of cyclopropane, divinyl ether, diethyl ether, trifluoroethyl vinyl ether, halothane, chloroform, trichloroethylene, and methoxyflurane between temperatures of 10 and 40°C and as a function of the blood hematocrit, Chiou and Hsiao [12] used the partition coefficient method to determine the degree of protein binding. A simple equation was derived to estimate the fraction of volatile or gaseous compounds bound to the protein in the aqueous protein solution. It is based on the assumption that an increase in the partition coefficient of the compound between the protein solution and air phase as compared to that between the aqueous phase not containing protein and the air phase is attributed to the binding of compound to the protein. The compounds specifically studied in this protein-binding investigation were fluorocarbons FCC 11, FCC 12, and FCC 114, halothane, and cyclopropane.

They derived the simple equation in the following manner: The percentage of a compound that is protein-bound in an aqueous protein solution at equilibrium is given by

$$\%\ \text{bound} = \frac{100C_B}{C_B + C_F} = \left(\frac{C_B}{C_T}\right)100 \tag{1.9}$$

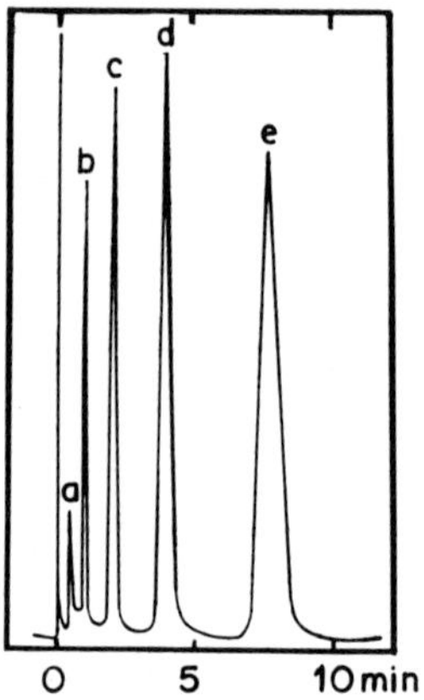

Figure 1.36. Gas chromatogram of a mixture of fluorochlorocarbons: (a) oxygen contamination; (b) FCC 12, 0.33 ng; (c) FCC 114, 4.05 ng; (d) FCC 11, 0.06 ng; (e) FCC 113, 1.26 ng. From Rauwns et al. [98], courtesy of Journal of Pharmacy and Pharmacology.

TABLE 1.15

Some Analytical Data of FCCs 11, 12, 113, and 114[a]

	FCC			
	11	12	113	114
Retention time (min)	4.0	1.0	7.5	2.0
Distribution coeff. Nitrogen: water[b]	0.53 ± 0.05	3.89 ± 0.28	0.96 ± 0.06	4.92 ± 0.38
Distribution coeff. Nitrogen: plasma[c]	0.51 ± 0.03	8.35 ± 0.75	-	4.15 ± 0.31
Detection limit[d] (pg)	0.5	3.3	10	35
Detection limit in sample[e] (ng ml^{-1} blood)	0.21	1.35	4.20	14.0

[a]From Rauwns et al. [98], courtesy of Journal of Pharmacy and Pharmacology.

[b]Mean ± S.D., room temperature (20°C).

[c]Rat plasma, mean ± S.D., room temperature (20°C).

[d]Sample 100 μl, attenuation 1 x 10^{-9}.

[e]Headspace sample 100 μl, attenuation 1 x 10^{-9}, gas:liquid ratio 40.

where

C_F = concentration of free or unbound compound
C_B = concentration of protein-bound compound
C_T = total compound concentration in the solution, $C_B + C_F$

Since the volatile or gaseous compound in a closed system will distribute between the protein solution and the headspace at equilibrium according to its partition coefficient, P_p, this can be expressed by

$$P_p = \frac{C_p}{C_g} \tag{1.10}$$

where C_p is the compound's concentration in the protein solution and C_g is the concentration of the compound in the headspace. By substituting C_T in Eq. (1.9) with C_p and rearranging, Eq. (1.9) becomes

$$\% \text{ bound} = \left(\frac{C_p - C_f}{C_p}\right)100 \tag{1.11}$$

Dividing Eq. (1.11) by C_g, one obtains

$$\% \text{ bound} = \frac{(C_p/C_g) - (C_f/C_g)}{C_p/C_g} \quad 100 \tag{1.12}$$

$$= \left(\frac{P_p - P_f}{P_p}\right)100 \tag{1.13}$$

The value of P_p can be determined experimentally using Eq. (1.10).

The recovery of fluorocarbon propellants from buffer and 5% protein solution with cyclohexane can be determined by injecting 5 to 10 μl of the cyclohexane phase onto the column of the gas chromatograph, whose operational parameters were essentially the same as those reported previously by Chiou and Niazi [139]. The peak height of the fluorocarbon from the cyclohexane extract is designated as PH_c.

The partition coefficients of the fluorocarbons in the buffer or protein solutions can be determined from the following expression:

$$P = \frac{C_l}{C_g} = \frac{(C_c)\,(V_c)/(f)\,(V_l)}{C_g} = \frac{(C_c)\,(V_c)}{(f)\,(C_g)\,(V_l)} \tag{1.14}$$

where

C_l = fluorocarbon concentration in buffer or protein solution at equilibrium
C_g = fluorocarbon concentration in the gaseous phase at equilibrium

C_c = fluorocarbon concentration in the cyclohexane after extraction
V_c = cyclohexane volume used for extraction
V_l = volume of buffer or protein solutions transferred for extraction
f = fraction of recovery

Since the fluorocarbon concentration is proportional to its peak height as indicated on the gas chromatogram, C_c and C_g can be substituted with PH_c and PH_g. Then

$$P = \frac{(PH_c)\ (V_c)}{(f)\ (PH_g)\ (V_l)} \tag{1.15}$$

The free form in the protein solution, which is thermodynamically in equilibrium with the compound in the headspace (present here only in the free form), can be shown schematically as follows:

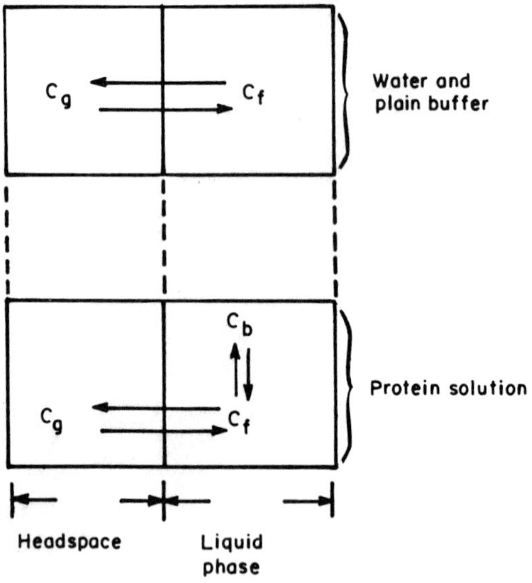

In this diagram, the partition coefficient (P_f) between the headspace and the free form in the protein solution can be determined directly in the same liquid system without the protein.

From their studies, it was estimated that 65.8, 51.2, and 73.3% of fluorocarbons 11, 12, and 114, respectively, were protein-bound in the 5% human albumin solution, and 71.2, 51.4, and 78.8% of them, respectively, were bound in the 5% bovine albumin solution. From the literature, it was also calculated that 20.3% of cyclopropane was bound in 5% human albumin solution and 63.6% of halothane was bound in 4% human albumin solution.

IV. STERILIZING AGENTS, COMMON ORGANIC SOLVENTS IN BLOOD, AND RIOT-CONTROL AEROSOL IRRITANTS

Widely used to sterilize surgical instruments, pharmaceutical and hospital equipment, pharmaceuticals, and such diverse materials as solid fuels, satellite components, and foods that would be damaged by excess heat [13, 14,140], ethylene oxide is highly flammable and must therefore be mixed with an inert gas to protect against combustion [140]. For this purpose, Freons and carbon dioxide are commonly used; sterilizing mixtures are available commercially in aerosol bombs. Since it is advisable to monitor the concentration of the components in order to avoid the possibility of explosion, this type of analysis can be easily and accurately performed by gas chromatography. If the instrument shown schematically in Figure 1.12 is used in its single-column mode (that is, placing an empty column in the column 1 position and a 12-ft x 3/16-in. aluminum column in the column 2 position, packed with 14% (w/w) dioctyl sebacate and 6% Ucon 50-HB-2000 on acid-base washed Chromosorb P), a mixture such as that shown in Figure 1.37 can be resolved. Because ethylene oxide is absorbed by Drierite dessiccant, the drying tube shown in Figure 1.12 can be filled with inert Teflon packing in order to circumvent this absorption problem.

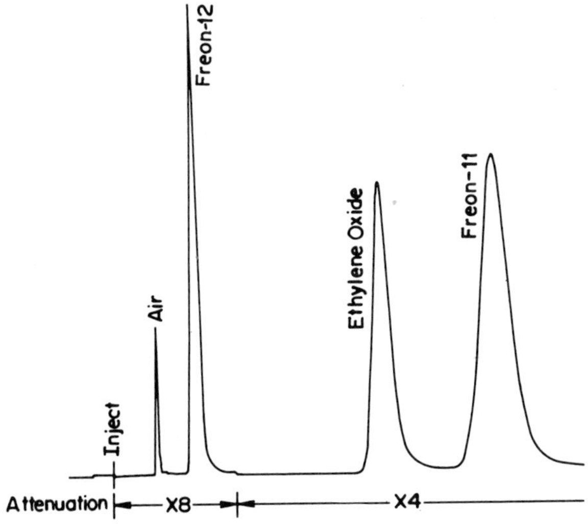

Figure 1.37. Analysis of a commercial sterilizing mixture with the Model 29 Gas Partitioner. Courtesy of Fisher Scientific Co.

Adler [13] determined the presence of ethylene oxide and ethylene glycol in ethylene oxide-sterilized pharmaceuticals (steroids, vitamins, antibiotics) using the following GC parameters: (1) a Perkin-Elmer 154-B with thermal conducitivity detector; (2) a 2-m x 1/4-in.-o.d. copper column filled with 25% dimethyl sulfolane coated on 30-60 mesh firebrick; 1-m and 2-m x 1/4-in. copper columns in series packed with 25%, 1,2,4-butanetriol and 25% PEG 1500, respectively; each coated on 30-60 mesh firebrick; (3) carrier-gas flow rate, 45 ml/min; (4) column temperature, 30°C for the dimethyl sulfolane and 40°C for the two columns connected in series; (5) internal standard, propylene oxide. With this proposed method, 2% precision is attainable when performing quantitative studies.

Spitz and Weinberger [141] simultaneously determined ethylene chlorohydrin and ethylene glycol in an aqueous solution under isothermal conditions using a single polyethylene glycol column, while ethylene oxide was examined more adequately with a styrene-divinylbenzene copolymer column. Because the demand for ethylene oxide as a sterilizing agent had stimulated a large amount of research on the possible toxicological effects of ethylene chlorohydrin (2-chloroethanol) and ethylene glycol, which are associated side products of this sterilant, a method was required for their determination in the nanogram range.

These quantitative requirements were met using a GC unit equipped with a dual-flame ionization detector (F&M model 5750). The column found most suited for the analysis of ethylene oxide was a coiled SS column, 6 ft x 1/8 in. i.d., containing a styrene-divinylbenzene copolymer resin (80-100 mesh, 300 to 400 m^2/g of surface area).

However, for the analysis of ethylene chlorohydrin and ethylene glycol, the following coiled glass column was used: 6 ft by 2 mm i.d., containing 3% polyethylene glycol coated on a styrene-divinylbenzene copolymer resin (80-100 mesh, less than 50 m^2/g surface area).

For the ethylene chlorohydrin and ethylene glycol analysis, the instrument was operated isothermally at a column temperature of 168°C, an injector temperature of 195°C, and a detector temperature of 220°C. The helium carrier-gas flow rate was maintained at 30 ml/min.

For the analysis of ethylene oxide, the following operational conditions were used: (1) column temperature, 100°C for 9 min, then the instrument was programmed at 50°C/min up to a maximum of 200°C and held for 4 min at this temperature; (2) injector temperature, 125°C; (3) detector temperature, 220°C; (4) helium carrier-gas flow rate, 30 ml/min.

Using the above conditions, the separation of ethylene chlorohydrin and ethylene glycol is rather good, with negligible baseline noise noted (Fig. 1.38). On the other hand, the excellent stability of the styrene-divinylbenzene resin coupled with its ability to resolve ethylene oxide from a solvent such as acetone (and its impurities) makes this packing very desirable, as noted in Figure 1.39. On this column, the approximate lower limit of detection for ethylene oxide was found to be about 0.7 ng/μl, using a 1-μl injection of acetone. For ethylene chlorohydrin and ethylene glycol, the detection limits were approximately 0.6 and 1.6 ng/μl, respectively.

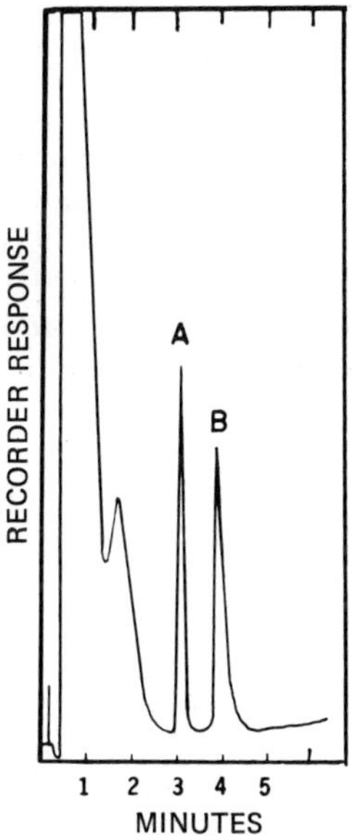

Figure 1.38. Typical gas chromatogram of 5.7 ng of ethylene chlorohydrin (A) and 6.4 ng of ethylene glycol (B). From Spitz and Weinberger [14], courtesy of Journal of Pharmaceutical Sciences.

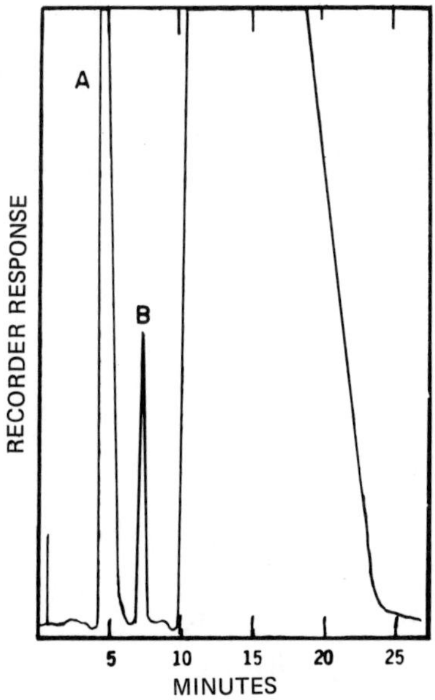

Figure 1.39. Typical gas chromatogram of 5.7 ng of ethylene oxide (B). (A) is an acetone impurity peak. From Spitz and Weinberger [14], courtesy of Journal of Pharmaceutical Sciences.

In addition to sterilizing agents, gas chromatography has been accepted as the method of choice for the analysis of microgram amounts of 34 fumigant gases used to control insects and fungi in fruit, textiles, tobacco, and dairy products and to combat plant diseases, nematodes, and anthrax organisms in soil. Berck [141] employed a 6-ft column packed with 10% SE-30 on Diatoport S, 60-80 mesh, and a thermal conductivity cell to investigate the following compounds: methyl chloride, methyl bromide, ethylene oxide, phosphene, hydrogen cyanide, acrylonitrile, methylene dichloride, carbon disulfide, chloroform, bromoform, chloropicrin, carbon tetrachloride, 1,1,1-trichloroethane, sym-tetrachloroethane, ethylene dichloride, ethylene dibromide, trichloroethylene, tetrachloroethylene, propylene dichloride, 1-bromopropane, 2-bromopropane, 1,3-dibromopropane, 3-bromopropane, 1-bromobutane, 2-bromobutane, 1-bromopentane, 2-bromopentane, chlorobenzene, bromobenzene, o-dichlorobenzene, p-dichlorobenzene, β,β'-dichloroethyl ether, and hexachlorobutadiene. Absolute and relative retention times of these fumigants are listed in Table 1.16, and Figure 1.40 is a typical chromatogram showing the elution of gas mixtures under isothermal and programmed temperature conditions.

TABLE 1.16

Absolute and Relative Retention Times of 34 Fumigants[a,b]

No.	Fumigant	Formula	Col. temp. (°C)	Abs. t_r (min)	Rel. t_r^c
1	Methyl chloride	CH_3Cl	60	0.84	0.71
2	Methyl bromide	CH_3Br	60	1.00	0.85
3	Ethylene oxide	$(CH_2)_2O$	80	0.85	0.89
4	Phosphine	PH_3	75	0.70	0.71
5	Hydrogen cyanide	HCN	75	0.88	1.11
6	Methylene dichloride	CH_2Cl_2	60	1.32	1.12
7	Carbon disulfide	CS_2	60	1.52	1.29
8	Chloroform	$CHCl_3$	60	2.22	1.89
9	1,1,1-Trichloroethane	$CCl_3 \cdot CH_3$	60	2.40	2.04
10	Ethylene dichloride	$CH_2Cl \cdot CH_2Cl$	60	2.67	2.27
11	Carbon tetrachloride	CCl_4	60	3.17	2.70
12	Propylene dichloride	$CH_2Cl \cdot CHCl \cdot CH_3$	60	3.55	3.02

TABLE 1.16 (continued)

No.	Fumigant	Formula	Col. temp. (°C)	Abs. t_r (min)	Rel. t_r^c
13	Ethyl bromide	C_2H_5Br	70	1.17	1.11
14	2-Bromopropane	$CH_3 \cdot CHBr \cdot CH_3$	70	1.55	1.48
15	3-Bromopropene	$CH_2:CH \cdot CH_2Br$	70	1.80	1.71
16	1-Bromopropane	C_3H_7Br	70	1.93	1.84
17	2-Bromobutane	$C_2H_5 \cdot CHBr \cdot CH_3$	70	2.63	2.51
18	Trichloroethylene	$CHCl:CCl_2$	70	2.74	2.61
19	Chloropicrin	$CCl_3 \cdot NO_2$	80	(3 peaks)	d
20	1-Bromobutane	$C_3H_7 \cdot CH_2Br$	80	2.58	2.71
21	1-Bromopentane	$C_4H_9 \cdot CH_2Br$	100	2.38	2.83
22	Ethylene dibromide	$CH_2Br \cdot CH_2Br$	100	2.57	3.05
23	Tetrachloroethylene	$CCl_2:CCl_2$	100	2.63	3.13
24	Acrylonitrile	$CH_2:CHCN$	110	0.92	1.21
25	2-Bromopentane	$C_3H_7 \cdot CHBr \cdot CH_3$	110	2.28	3.02
26	Chlorobenzene	C_6H_5Cl	125	1.88	2.64
27	Bromoform	$CHBr_3$	125	2.33	3.28
28	sym-Tetrachloroethane	$CHCl_2 \cdot CHCl_2$	125	2.57	3.70
29	1,3-Dibromopropane	$CH_2Br \cdot CH_2 \cdot CH_2Br$	125	2.78	3.90
30	Bromobenzene	C_6H_5Br	125	2.85	4.00
31	β,β'-Dichloroethyl ether	$(C_2H_4Cl)_2O$	125	3.03	4.26
32	p-Dichlorobenzene	$C_6H_4Cl_2$	125	4.05	5.68
33	o-Dichlorobenzene	$C_6H_4Cl_2$	125	4.20	5.90
34	Hexachlorobutadiene	$CCl_2:CCl \cdot CCl:CCl_2$	180	2.68	4.32

[a]From Berck [141], courtesy of Journal of Agricultural and Food Chemistry.

[b]2 μg of fumigant; instrument attenuation = 1 x; flow rate = 50 cm^3 helium/min.

[c]Relative to n-pentane = 1.00.

[d]Three peaks for chloropicrin at 0.90, 2.75, and 3.62 min abs. t_r.

t_r = retention time.

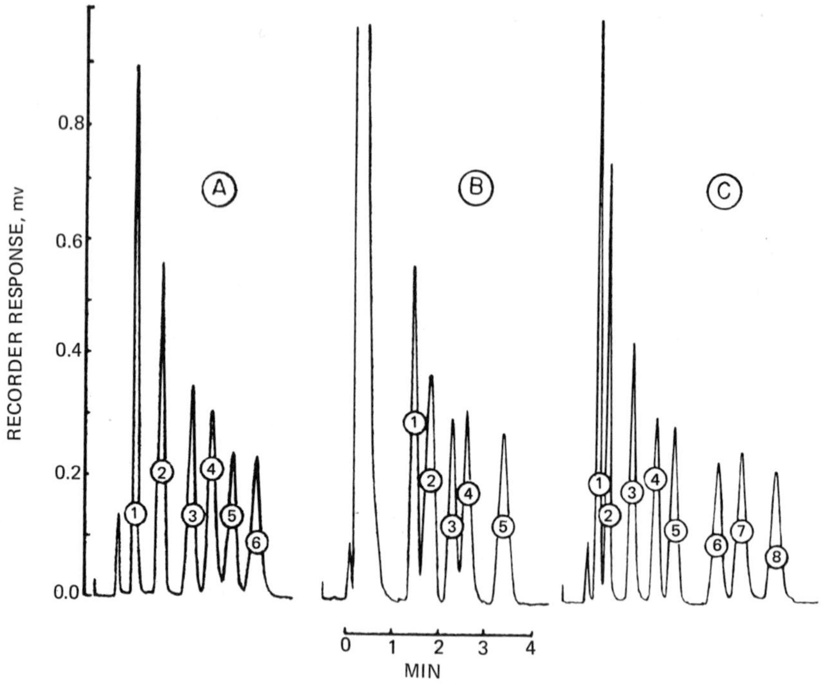

Figure 1.40. Elution of gas mixtures under isothermal and programmed temperature conditions. (A) Isothermal separation of six gases at 60°C, attenuation 1 x, abs. t_r (minutes). (1) methylbromide, 2.5 μg, 1.02 min; (2) carbon disulfide, 4 μg, 1.55 min; (3) chloroform, 4 μg, 2.23 min; (4) ethylene dichloride, 4 μg, 2.65 min; (5) carbon tetrachloride, 2 μg, 3.13 min; (6) propylene dichloride, 3 μg, 3.58 min. (B) Isothermal separation of five gases at 125°C, attenuation 1 x, abs. t_r (minutes); (1) chlorobenzene, 4 μg, 1.97 min; (2) bromoform, 4 μg, 2.37 min; (3) sym-tetrachloroethane, 3 μg, 2.72 min; (4) β,β'-dichloroethyl ether, 3 μg, 3.10 min; (5) p-dichlorobenzene, 4 μg, 4.02 min. (C) PTGC separation of eight gases, started at 50°C, column temperature and attenuation 1 x. After air peak (0.68 min), programmed at 7.9°C/min; (1) methyl chloride, 3 μg, 0.90 min; (2) methyl bromide, 2 μg, 1.10 min; (3) methylene dichloride, 3 μg, 1.65 min; (4) 2-bromopropane, 3 μg, 2.20 min; (5) chloroform, 3 μg, 2.60 min; (6) carbon tetrachloride, 2 μg, 3.53 min; (7) propylene dichloride, 3 μg, 4.05 min; (8) 1-bromobutane, 2 μg, 4.80 min. From Berck [141], courtesy of Journal of Agricultural and Food Chemistry.

Berck [142] also utilized gas chromatography to determine the sorption by 51 cereal products of ethylene dibromide, ethylene dichloride, and carbon tetrachloride applied in the vapor phase singly and in various admixtures for 5 hr at 27°C in microfumigation chambers. The resulting data showed that the order of sorption by most of the substrates was ethylene dibromide > ethylene dichloride > carbon tetrachloride. The uptake of the ethylene compounds increased with an increase in the moisture content of wheat and with a decrease in particle size by grinding.

For the analysis of nematocide residues, Beckman and Bevenue [143] employed an electron-capture gas chromatographic unit for the determination of 1,2-dibromo-3-chloropropane and tetrachlorothiaphene; the separation was performed with a 5-ft x 1/8-in.-i.d. SS column packed with 5% SE-30 silicone gum rubber on 60-80 mesh Chromosorb W. Whereas only as little as 0.1 μg of tetrachlorothiaphene could be detected by using a chromatographic system equipped with a microcoulometer detector, a 10% scale deflection on the recorder with an electron-affinity detector was observed with 15 pg (1 pg = 1×10^{-6} μg).

Cullen and Stanovick [144] developed a method for determining Lanstan (1-chloro-2-nitropropane) residues on agricultural crops by electron-affinity detection. The column was a 4-ft x 1/4-in. aluminum tube packed with 4% XE-60 on 80 mesh Chromosorb W. At a column temperature of 80°C, Lanstan was eluted in 3 min with excellent detector response. With a special extraction-cleanup procedure, the method could easily detect 10 ppb (10 pg). Typical recovery data generally ranged between 75 and 105% after fortifications of 10 to 110 ppb were made.

Premel-Cabic et al. [15] described a method for the determination of small amounts of 15 organic solvents in blood. After equilibration of the blood with air in a tightly closed tube, air is injected into the chromatograph. Two columns were used with a flame ionization detector which is sensitive to hydrocarbons and halogenated compounds. Using an internal standard, the sensitivity (1.5 mg/liter for trichloroethylene) was sufficient for the needs of antipoison centers.

As previously noted, two columns were used, each 1.5-m x 2 mm i.d., one filled with 20% Carbowax 20M coated on Spherosil XOB 75 (65-150 mesh) and maintained at 90°C at a nitrogen carrier-gas flow rate of 17 ml/min, the other packed with 120-150 mesh Porapak Q and operated at 190°C with a carrier-gas flow rate of 23 ml/min. Using either toluene or trichloroethylene as the internal standard, typical chromatographic separations of a 15-component mixture on both the Carbowax and Porapak Q columns are shown in Figures 1.41 and 1.42, respectively. Compound identification is based on its retention time relative to toluene on either column used (Table 1.17).

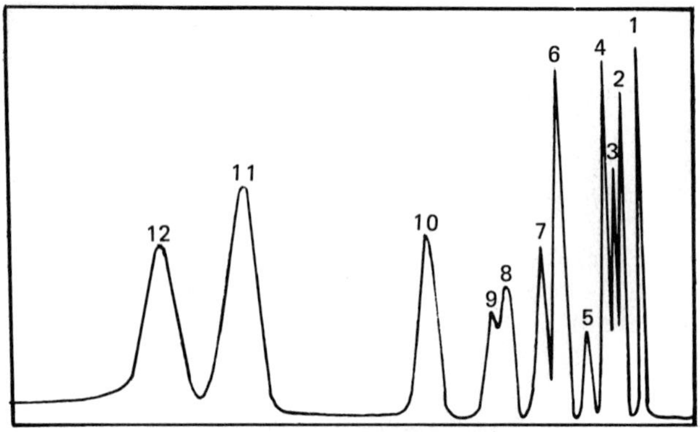

Figure 1.41. Chromatogram of a mixture of 15 solvents: 1, pentane; 2, dichloroethylene and ether; 3, methylene chloride; 4, hexane; 5, carbon tetrachloride; 6, benzene and trichloroethane; 7, trichloroethylene and chloroform; 8, dichloroethane; 9, tetrachloroethylene; 10, toluene; 11, chlorobenzene; 12, m-xylene. (Carbowax 20M column.) From Premel-Cabic et al. [15], courtesy of Clinica Chemica Acta.

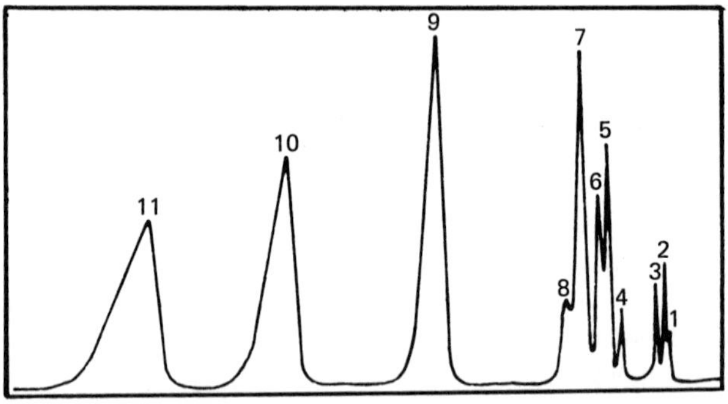

Figure 1.42. Chromatogram of a mixture of 15 solvents: 1, pentane; 2, methylene chloride and ether; 3, dichloroethylene; 4, chloroform; 5, hexane; 6, dichloroethane, trichloroethane, and carbon tetrachloride; 7, benzene; 8, trichloroethylene; 9, tetrachloroethylene and toluene; 10, chlorobenzene; 11, m-xylene. (Porapak Q column at 190°C.) From Premel-Cabic et al. [15], courtesy of Clinica Chemica Acta.

TABLE 1.17

Relative Retention Times of 15 Solvents (Toluene = 1.00)[a]

Solvent	Relative retention time	
	Carbowax 20M	Porapak Q
Pentane	0.19	0.18
Ether	0.26	0.19
Dichloroethylene	0.26	0.22
Methylene chloride	0.29	0.19
Hexane	0.32	0.39
Carbon tetrachloride	0.40	0.42
Trichloroethane	0.49	0.42
Benzene	0.49	0.49
Trichloroethylene	0.56	0.55
Chloroform	0.56	0.35
Dichloroethane	0.69	0.42
Tetrachloroethylene	0.75	1.00
Toluene	1.00	1.00
Chlorobenzene	1.70	1.54
m-Xylene	2.00	2.03

[a]From Premel-Cabic et al. [15], courtesy of Clinica Chemica Acta.

The analysis of irritants or riot-control agents is also amenable to detection by GC as well as GC-MS instrumentation. These agents provide military and police forces, custodial officers, and individuals with alternatives to traditional weapons.

Sass et al. [18] described GC procedures applicable to the estimation of α-bromobenzylnitrile (bromobenzylcyanide, CA), o-chlorobenzalmalonitrile (CS), and α-chloroacetophenone (CN) as concentrated and dilute solutions and when sampled directly as vapor or aerosol.

Using an F&M model 810 gas chromatograph equipped with a thermal conducitivity, flame ionization, or electron-capture detector, CA, CN, or CS purity (with 1,10-dibromodecane as internal standard) was determined by injecting aliquots of benzene solution containing known sample weights (about 50 mg/ml of benzene) onto a 5.5-ft x 1/4-in.-o.d. borosilicate glass column packed with 10% QF-1 on 60-80 mesh Gas-Chrom Q. With a carrier-gas (helium) flow rate of 90 ml/min and temperature programming from 65 to 200°C at 6°C/min, the retention times for CA, benzylcyanide, and internal standard were 15 (153°C), 12 (137°C), and 20 (185°C) min, respectively. On the basis of 16 separate samples involving more than 50 individual sample injections, a standard deviation of 1.0% was calculated.

For determining CN purity, CN was eluted in 14.5 min at 149°C while acetophenone had a retention time of 9 min at 119°C. For establishing CS purity, mixtures of CS and its two impurities (o-chlorobenzaldehyde and malonitrile) yielded elution times of 22 (193°C), 7.5 (110°C), and 7.0 (101°C) min, respectively.

Relative to the internal standard, the relative response factor, using the thermal conductivity detector for CA, CN, and CS were 1.030, 1.176, and 1.111, respectively.

To detect irritants (CA, CN, and CS) in microgram quantities, identical columns and conditions as those described for thermal conductivity studies were used, but in this situation, the flame ionization detector was fitted to the chromatograph. At the 10- to 200-μg/ml level for all three irritants, the minimum detectable quantity was 2 μg while the overall accuracy was ±2.0%.

On the other hand, trace determinations obtained via the electron-capture detector indicated that solutions containing CA in quantities as low as 0.2 μg/ml could be analyzed. In fact, the minimum detectable amount of CA by this method was about 4 ng (4×10^{-9} g). At a carrier gas (90% argon-10% methane mixture) flow rate of 85 ml/min, the retention time of CA at a column temperature of 130°C was 6.5 min. Using isothermal column temperatures of 130 and 150°C for CN and CS, respectively, the minimum detectable amount for CN and CS was 0.2 ng (2×10^{-9} g) and 0.1 ng, respectively.

The systems for "macro" to trace analysis of CA, CN, and CS were designed to fit the most difficult of the three compounds, namely CA. The compound was known to pyrolyze readily in the presence of metals to form stilbene-type and a variety of other products. To preclude metal-catalyzed decomposition, all-glass systems were used wherever possible.

Zerba and Ruveda [17] chromatographed the following riot-control agents using a Varian Aerograph 1522-B temperature-programmed chromatograph equipped with dual-flame detectors and a 6-ft glass column packed with OV-17 coated on Varaport 30, 100-120 mesh: benzyl bromide, xylyl bromide (mixture of isomers), α-chloroacetophenone (CN), o-chlorobenzalmalonitrile (CS), and diphenylaminechloroarsine (adamsite), including such important impurities as diphenylamine and bromotoluene (mixture of o- and p-isomers). The instrument was operated at the following experimental conditions: inlet temperature, 200°C; detector temperature, 240°C; column temperature, programmed at 12°C/min from 55 to 210°C; nitrogen carrier flow, 25 ml/min; hydrogen flow, 300 ml/min; air flow, 300 ml/min. The relative retention times as well as the elution temperatures of the compounds tested are given in Table 1.18 relative to α-chloroacetophenone. Adamsite was not detected under the experimental conditions used. Besides the possibility of identifying or determining the concentration of any of the riot-control agents, the procedure was found to be adequate for the separation of any mixture of them. Figure 1.43 shows a typical chromatogram of these compounds and their usual impurities.

Finally, Casselman et al. [16] developed a convenient method for the sampling and total analysis of the propellants and active agents contained in

TABLE 1.18

Relative Retention Times and Elution Temperatures of Riot-control Agents and Related Compounds[a]

Compound	Relative retention time[b]	Elution temperature (°C)
o- and p-Bromotoluene	0.64	91
Benzyl bromide	0.74	106
Xylyl bromide	0.92	131
α-Chloroacetophenone	1.00	143
o-Chlorobenzalmalononitrile	1.27	182
Diphenylamine	1.34	192

[a]From Zerba and Ruveda [17], courtesy of Journal of Chromatography.
[b]Relative to α-chloroacetophenone.

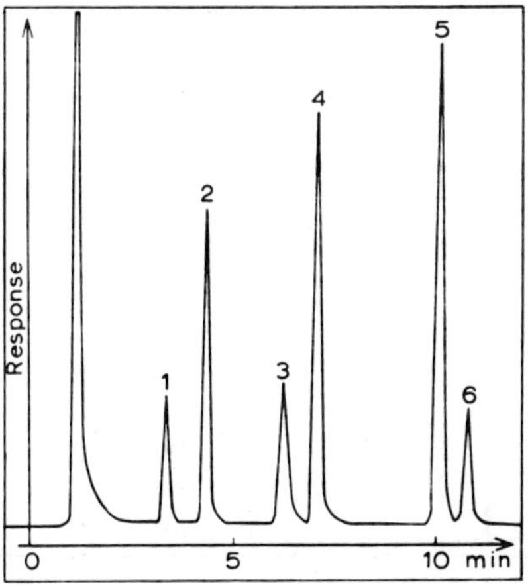

Figure 1.43. Chromatogram of the mixture of bromotoluene (1), benzyl bromide (2), xylyl bromide (3), α-chloroacetophenone (4), o-chlorobenzalmalononitrile (5), and diphenylamine (6). From Zerba and Ruveda [17], courtesy of Journal of Chromatography.

aerosol irritant projectors using a gas chromatograph (F&M 5750 with both a TC and FI detector) equipped with two columns: (1) a 6-ft x 1/8-in.-o.d. SS column packed with Porapak Q, 80-100 mesh, and (2) a 15-ft x 1/8-in.-o.d. SS column filled with 6% Carbowax 20M on Chromosorb G, AW- and DMCS-treated (80-100 mesh). The injection port and detector (TCD and FID) were kept at 170 and 190°C, respectively, whereas the helium carrier-gas flow rates were set at 25 ml/min and 30 ml/min, depending on whether the Porapak Q or Carbowax 20M was involved, with the fuel gases (FID), hydrogen and air, maintained at 10 and 30 psi, respectively. The Porapak Q column was operated isothermally at 170°C; the Carbowax 20M column required temperature programming from 60 to 180°C at 8°C/min.

The qualitative identification of all components other than the propellant was accomplished by coupling a fast-scanning Atlas CH4 mass spectrometer directly to the gas chromatograph at the exit port of the TCD detector. The coupling was made with a 1/16-in.-o.d. SS tube and a splitter valve. Pressure was held at 5×10^{-6} torr, ionizing voltage was 70 eV, and the ion source temperature was 250°C.

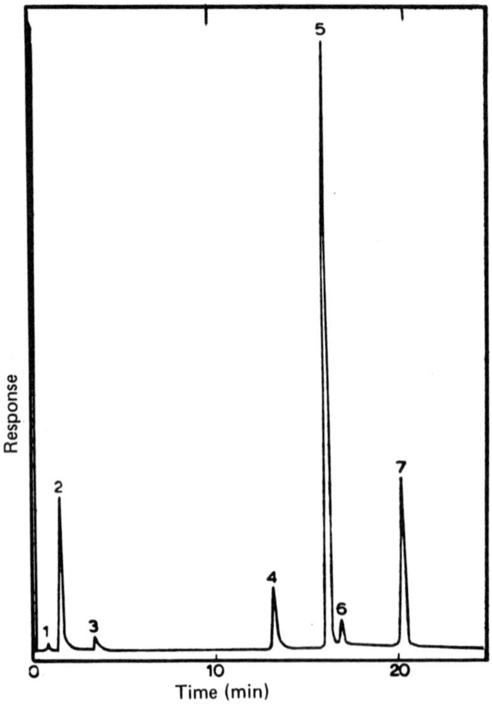

Figure 1.44. Chromatogram of the residual oil left from an aerosol tear gas sample. 1 = Dichlorodifluoromethane; 2 = trichlorofluoromethane; 3 = acetone; 4 = chloroacetone; 5 = bromoacetone; 6 = 1,1-dichloroacetone; 7 = 1,1-dibromoacetone. From Casselman et al. [16], courtesy of Journal of Chromatography.

Using a specially designed Pyrex bomb for sample collection, residual oil samples were injected into the integrated GC-MS instrument (Carbowax 20M column) to provide a qualitative analysis of the individual components. Figure 1.44 illustrates the chromatographic separation achieved; the retention times for dichlorodifluoromethane, trichlorofluoromethane, acetone, chloroacetone, bromoacetone, 1,1-dichloroacetone, and 1,1-dibromoacetone were 0.8, 1.6, 3.4, 13.2, 16.0, 16.8, and 20.1 min, respectively. The Porapak Q column [145] was used for the separation and identification of the two propellants (dichlorodifluoromethane and trichlorofluoromethane) due to the excellent separation which it afforded to these rather similar compounds. For an overall analysis, the propellants accounted for 97.7% of the original mixture, the main lachrymator, bromoacetone, being present to the extent of 1.5% and the remaining 0.8% of the mixture consisting of acetone, chloroacetone, 1,1-dibromoacetone and 1,1-dichloroacetone (these latter two components possibly present as deliberate additives or as impurities from the corresponding monohalogenated compounds).

REFERENCES

1. Hill, D. W., in R. P. W. Scott (Ed.), Preprints of the Third Symposium on Gas Chromatography, Butterworths, London, 1960.
2. Onchi, Y. and Asao, Y., Brit. J. Anesthesia, 33, 544 (1961).
3. Noehren, T. H. and Cudmore, J. W., Anesthesiology, 22, 519 (1961).
4. Butler, R. A. and Hill, D. W., Nature, 189, 488 (1961).
5. Adlard, E. R. and Hill, D. W., Nature, 186, 1045 (1960).
6. Gudzinowicz, B. J., Applications of Gas Chromatography to Pharceutical, Medical and Toxocological Analysis, Jarrell-Ash, Waltham, Mass., 1964.
7. Swinnerton, J. W., Linnenbom, V. J., and Cheek, C. H., Anal. Chem., 34, 483 (1962).
8. Purchase, I. F. H., Nature, 198, 895 (1963).
9. Curry, A. S., Hurst, G., Kent, N. R., and Powell, H., Nature, 195, 603 (1962).
10. Lowe, H. J. and Hagler, K., in R. Porter (Ed.), Gas Chromatography in Biology and Medicine, Churchill, London, 1969.
11. Chiou, W. L., Res. Commun. Clin. Pathol. Pharmacol., 7, 679 (1974).
12. Chiou, W. L. and Hsiao, J. H., Res. Commun. Clin. Pathol. Pharmacol., 8, 273 (1974).
13. Adler, N., J. Pharm. Sci., 54, 735 (1965).
14. Spitz, H. D. and Weinberger, J., J. Pharm. Sci., 60, 271 (1971).
15. Premel-Cabic, A., Cailleux, A., and Allain, P., Clin. Chim. Acta, 56, 5 (1974).

16. Casselman, A. A., Bannard, R. A. B., and Pottie, R. F., J. Chromatogr., 80, 155 (1973).
17. Zerba, E. N. and Ruveda, M. A., J. Chromatogr., 68, 245 (1972).
18. Sass, S., Fisher, T. L., Jascot, M. J., and Herban, J., Anal. Chem., 43, 462 (1971).
19. Smith, R. N., Swinehart, J., and Lesnini, D. G., Anal. Chem., 30, 1217 (1958).
20. Welton, W. M. and Drake, N. L., Ind. Eng. Chem., Anal. Ed., 1, 20 (1929).
21. Shah, M. S. and Oza, T. M., J. Chem. Soc., 32 (1931).
22. Smith, R. N., Lesnini, D. G., and Mooi, J., J. Phys. Chem., 60, 1063 (1956).
23. Szulczewski, D. H. and Higuchi, T., Anal. Chem., 29, 1541 (1957).
24. Boer, H., in D. H. Desty (Ed.), Vapor Phase Chromatography, Butterworths, London, 1957.
25. Work of R. L. Johnson, reported in A. C. De Graff, Jr., J. Gas Chromatogr., 2, 357 (1964).
26. Jay, B. E. and Wilson, R. H., J. Appl. Physiol., 15, 298 (1960).
27. Hamilton, L. H., Ann. N.Y. Acad. Sci., 102, 15 (1962-1963).
28. Hamilton, L. H., Smith, J. R., and Kory, R. C., AMA Annual Meeting, New York, June 1961.
29. Hamilton, L. H. and Kory, R. C., J. Appl. Physiol., 15, 829 (1960).
30. Forster, R. E., Fowler, W. S., and Bates, D. V., J. Clin. Invest., 33, 1128 (1954).
31. Riley, R. L., Shepard, R. H., Cohen, J. E., and Carroll, D. G., J. Appl. Physiol., 10, 573 (1954).
32. Lee, G. de J. and Du Bois, A. B., J. Clin. Invest., 34, 1380 (1955).
33. Grollman, A., The Cardiac Output of Man in Health and Disease, Thomas, Baltimore, 1932.
34. Roughton, F. J. W. and Forster, R. E., J. Appl. Physiol., 11, 290 (1957).
35. Roughton, F. J. W., Amer. J. Physiol., 143, 621 (1945).
36. Van Slyke, D. D. and Sendroy, J. (Tr.), J. Biol. Chem., 95, 509 (1932).
37. Hammarstrand, K., Varian Inst. Applic., 8 (1), 1974.
38. Patzelova, V., Chromatographia, 4, 174 (1971).
39. Hollis, O. L., Anal. Chem., 38, 309 (1966).
40. Hollis, O. L. and Hayes, W. V., J. Gas Chromatogr., 4, 235 (1966).
41. Dave, S. B., J. Chromatogr. Sci., 7, 389 (1969).
42. Hollis, O. L., J. Chromatogr. Sci., 11, 335 (1973).
43. Hirsch, R. F., Stober, H. C., Kowblansky, M., Hubner, F. N., and O'Connell, A. W., Anal. Chem., 45, 2100 (1973).
44. Ohzeki, K. and Kambarat, T., J. Chromatogr., 55, 319 (1971).
45. Czubryt, J. J. and Gesser, H. D., J. Gas Chromatogr., 6, 41 (1968).
46. Burnham, A. K., Calder, G. V., Fritz, J. S., Junk, G. A., Svec, H. J., and Willis, R., Anal. Chem., 44, 139 (1972).

47. Amberlite XAD Macroreticular Adsorbents, Rohm and Haas Co., Philadelphia, n.d.
48. Fritz, J. S. and Chang, R. C., Anal. Chem., 46, 938 (1974).
49. Wilson, R. H., Jay, B., Doty, V., Pingree, H., and Higgins, E., J. Appl. Physiol., 16, 374 (1961).
50. Galla, S. J. and Ottenstein, D. M., Ann. N.Y. Acad. Sci., 102, 4 (1962-1963).
51. Wilson, R. H., J. Gas Chromatogr., 2, 365 (1964).
52. Yee, H. Y., Anal. Chem., 37, 924 (1965).
53. Farhi, L. E., Edwards, A. W. T., and Homma, T., J. Appl. Physiol., 18, 97 (1963).
54. Ikels, K. G., J. Gas Chromatogr., 3, 359 (1965).
55. Natelson, S. and Stellate, R. L., Anal. Chem., 35, 847 (1963).
56. Ramsey, L. H., Science, 129, 900 (1959).
57. Lukas, D. S. and Ayres, S. M., J. Appl. Physiol., 16, 371 (1961).
58. Russ, C., Lee, J., Carr, Z., and Taylor, B., J. Gas Chromatogr., 2, 18 (1964).
59. Blackmore, D. J., in R. Porter (Ed.), Gas Chromatography in Biology and Medicine, Churchill, London, 1969.
60. Guild, L. V., Lloyd, M. I., and Aul, F., paper presented at ISA 2nd International Symposium on Gas Chromatography, East Lansing, Mich., June 1959.
61. Chambliss, K. W. and Nouse, D. C., Clin. Chem., 8, 654 (1962).
62. Thomas, D. J., Ching, N. P., and Nelson, T. F., Jr., Fed. Proc., 22, Part 1, 573 (1963).
63. Johns, T. and Thompson, B., The Analyzer, 4 (2), 13 (1963), Beckman Instruments, Inc., Fullerton, Calif.
64. Dominguez, A. M., Christenson, H. E., Goldbaum, L. R., and Stembridge, V. A., Toxicol. Appl. Pharmacol., 1, 135 (1959).
65. Wilson, R. H. and Jay, B. E., Clin. Res., 8, 92 (1960).
66. Natelson, S., Amer. J. Clin. Pathol., 21, 1153 (1951).
67. Klocke, F. J. and Rahn, H., J. Clin. Invest., 40, 279 (1961).
68. Van Slyke, D. D., Dillon, R. T., and Margaria, R., J. Biol. Chem., 105, 571 (1934).
69. Natelson, S. and Menning, C., Clin. Chem., 1, 165 (1955).
70. Natelson, S., Leighton, D., and Calas, C., Microchem. J., 6, 539 (1962).
71. Natelson, S. and Sheid, B., Clin. Chem., 7, 115 (1961).
72. Peters, J. P. and Van Slyke, D. D., Quantitative Clinical Chemistry (Methods), Vol. 2, Williams & Wilkins, Baltimore, 1956.
73. Goldbaum, L. R., Schloegel, E. L., and Dominguez, A. M., in A. Stolman (Ed.), Progress in Chemical Toxicology, Vol. 1, Academic Press, New York, 1963.
74. Hall, K. D., Norris, F., and Downs, S., Anesthesiology, 21, 522 (1960).

75. Summers, F. W. and Adriani, J., Anesthesiology, 22, 100 (1961).
76. Baker, J. M. and King, B. D., Anesthesiology, 24, 122 (1962).
77. Zauder, H. L. and Orkin, L. R., Fed. Proc., 21, 328 (1962).
78. Cheneau, M. and Guichon, G., Bull. Soc. Chim. Fr., 2601 (1963).
79. Zauder, H. L. and Orkin, L. R., Can. Anaesth. Soc. J., 10, 228 (1963).
80. Northington, M. S. and Owens, G., Anal. Biochem., 9, 48 (1964).
81. Lowe, H. J., J. Gas Chromatogr., 2, 380 (1964).
82. Gadsen, R. H. and McCord, W. M., J. Gas Chromatogr., 2, 7 (1964).
83. Lowe, H. J. and Beckham, L. M., in H. A. Szymanski (Ed.), Biomedical Applications of Gas Chromatography, Plenum Press, New York, 1964.
84. Cohen, E. N., Bellville, J. W., Budzikiewicz, H., and Williams, D. H., Science, 141, 899 (1963).
85. Cohen, E. N. and Brewer, H. W., J. Gas Chromatogr., 2, 261 (1964).
86. Chenoweth, M. B. and Brewer, H. W., in B. R. Fink (Ed.), Toxicity of Anesthetics, Williams & Wilkins, Baltimore, 1968.
87. Butler, R. A. and Linde, H. W., Anesthesiology, 25, 397 (1964).
88. Phillips, O. C., Anesthesiology, 26, 104 (1965).
89. Cohen, E. N., Brewer, H. W., Bellville, J. W., and Sher, R., Anesthesiology, 26, 140 (1965).
90. Rutledge, C. O., Seifen, E., Alper, M. H., and Flacke, W., Anesthesiology, 24, 862 (1963).
91. Theye, R. A., Anesthesiology, 25, 75 (1964).
92. Epstein, R. M., Rackow, H., Salanitre, E., and Wolf, G. L., Anesthesiology, 25, 364 (1964).
93. Stewart, R. D., Swank, J. D., Roberts, C. B., and Dodd, H. C., Nature, 198, 696 (1963).
94. Brachet-Liermain, A., Ferrus, L., and Caroff, J., J. Chromatogr. Sci., 9, 49 (1971).
95. Smith, R. V., Amer. Lab., pp. 58-66, November 1971.
96. Jacobs, E. S., Anesthesia & Analgesia, 43, 177 (1964).
97. Gadsen, R. H., McCord, W. M., Woods, E. F., and Bagwell, E. E., Anesthesiology, 23, 831 (1962).
98. Rauwns, A. G., Olling, M., and Wibowo, A. E., J. Pharm. Pharmacol., 25, 718 (1973).
99. Fabian, L. W., Dewitt, H., and Carnes, M. A., Anesthesia & Analgesia, 39, 456 (1960).
100. Gloesner, E., J. Pharm. Belg., 13, 585 (1958).
101. Gjaldbaek, J., Dansk. Tidsskr. Farm., 33, 158 (1959).
102. Lowe, H. J., F&M Scientific Corp. Facts and Methods, 3, 5 (1963).
103. Butler, R. A. and Freeman, J., Brit. J. Anaesth., 34, 440 (1962).

104. Dyfverman, A. and Sjovall, J., Acta Anesth. Scand., 6, 171 (1962).
105. Gadsen, R. H., Risinger, B. H., and Bagwell, E. E., Can. Anaesth. Soc. J., 12, 90 (1965).
106. Van Dyke, R. A., Chenoweth, M. B., and Larson, E. R., Nature, 204, 471 (1964).
107. Gadsen, R. H. and McCord, W. M., Fed. Proc., 21, 327 (1962).
108. Lowe, H. J., Anesthesiology, 25, 808 (1964).
109. Koudela, S. and Lukaci, J., Soudni Lekarstvi, 10, 8 (1965).
110. Rehder, K., Forbes, J., Hessler, O., and Gossmann, K., Anesthesist, 15, 162 (1966).
111. Wortley, D. J., Herbert, P., Thornton, J. A., and Whelpton, D., Brit. J. Anaesth., 40, 624 (1968).
112. Tiengo, M., Ann. Obstet. Gynecol., 89, 543 (1967).
113. Albin, M. S., Horrocks, L. A., and Kretchmer, H. E., Anesthesiology, 25, 672 (1964).
114. Gjaldbaek, J. C. and Worm, K., Dansk. Tidsskr. Farm., 39, 141 (1965).
115. Wolfson, B., Ciccarelli, H. E., and Siker, E. S., Brit. J. Anaesth., 38, 591 (1966).
116. Chapman, J., Hill, R., Muir, J., Suckling, C. W., and Viney, D. J., J. Pharm. Pharmacol., 19, 231 (1967).
117. Davies, D. D. and Mathias, J. A., Brit. J. Pharmacol., 40, 596P (1970).
118. Scipioni, A., Gambaretto, G., Triolo, G., and Fraccaro, C., Chem. Ind. (Milan), 49, 577 (1967).
119. Douglas, R., Hill, D. W., and Wood, D. G. L., Brit. J. Anaesth., 42, 119 (1970).
120. Herbert, P., J. Med. Lab. Technol. (London), 25, 233 (1968).
121. Scherer, O. and Weigand, W., Anesthesist, 13, 313 (1964).
122. Butler, R. A., Kelly, A. B., and Zapp. J., Anesthesiology, 28, 760 (1967).
123. Yokota, T., Hitomi, Y., Ohta, K., and Kosaka, F., Anesthesiology, 28, 1064 (1967).
124. Laasberg, L. H. and Hedley-Whyte, J., Anesthesiology, 32, 351 (1970).
125. Fink, B. R. and Morikawa, K., Anesthesiology, 32, 451 (1970).
126. Rackow, H., Salanitre, E., Epstein, R. M., Wolf, G. L., and Perl, W., J. Appl. Physiol., 20, 611 (1965).
127. Harley, J., Nel, W., and Pretorius, V., Nature, 181, 177 (1958).
128. Phillips, T. R. and Owens, D. R., in R. P. W. Scott (Ed.), Preprints of the Third Symposium on Gas Chromatography, Butterworths, London, 1960.
129. Cadman, W. J. and Johns, T., Pittsburgh Conference on Analytical Chemistry and Applied Spectroscopy, March 1958.

130. Bunker, J. P. and Blumenfeld, C. M., New Engl. J. Med., 268, 531 (1963).
131. Lu, G., Johnson, S. L., Ling, M. S., and Krantz, J. C., Anesthiology, 14, 466 (1953).
132. Clayton, J. W., J. Occupat. Med., 4, 262 (1962).
133. Duncan, W. A. M. and Raventos, J., Brit. J. Anaesth., 31, 302 (1959).
134. Chenoweth, M. B., Robertson, D. N., Erley, D. S., and Golhke, R., Anesthesiology, 23, 101 (1962).
135. Van Stekelenburg, G. J. and De Bruijn, J. W., Clin. Chim. Acta, 28, 233 (1970).
136. Dollery, C. T., Draffan, G. H., Davies, D. S., Williams, F. M., and Conolly, M. E., Lancet, 2, 1164 (1970).
137. Paterson, J. W., Sudlow, M. F., and Walker, S. R., Lancet, 2, 565 (1971).
138. Shargel, L. and Koss, R., J. Pharm. Sci., 61, 1445 (1972).
139. Chiou, W. L. and Niazi, S., Res. Commun. Clin. Pathol. Pharmacol., 6, 481 (1973).
140. Fisher Gas Partitioner Brochure, Fisher Scientific Co., Pittsburgh.
141. Berck, B., J. Agr. Food Chem., 13, 373 (1965).
142. Berck, B., J. Agr. Food Chem., 13, 248 (1965).
143. Beckman, H. and Bevenue, A., J. Agr. Food Chem., 11, 479 (1963).
144. Cullen, T. E. and Stanovick, R. P., Niagara Chemical Division, FMC Corp., Middleport, N. Y., private communication, February 1965.
145. Foris, A. and Lehman, J. G., Separation Sci., 4, 230 (1969).

GENERAL REFERENCES/REVIEWS FOR GC/MS DRUG ANALYSIS

A. Gas-Liquid Chromatography

1. Dal Nogare, S. and Juvet, R. S., Jr., Gas-Liquid Chromatography, Wiley, New York, 1962.
2. Keulemans, A. I. M., Gas Chromatography, Reinhold, New York, 1959.
3. Kaiser, R., Gas Phase Chromatography, Vols. I-III, Butterworths, Washington, 1963.
4. Ettre, L. S. and Zlatkis, (Eds.), The Practice of Chromatography, Wiley, New York, 1967.
5. Gudzinowicz, B. J., Gudzinowicz, M. J., and Martin, H. F., Fundamentals of Integrated GC-MS, Part I, Marcel Dekker, New York, 1976.

6. Jones, R. A., An Introduction to Gas-Liquid Chromatography, Academic Press, New York, 1970.
7. Harris, W. E. and Habgood, H. W., Programmed Temperature Gas Chromatography, Wiley, New York, 1966.
8. Ettre, L. S., Open Tubular Columns, Plenum Press, New York, 1965.
9. Littlewood, A. B., Gas Chromatography, 2nd ed., Academic Press, New York, 1970.
10. Purnell, J. H., Gas Chromatography, Wiley, New York, 1962.
11. Giddings, J. C., Grushka, E., Keller, R. A., and Cazes, J. (Eds.), Advances in Chromatography, Vols. I-XII, Marcel Dekker, New York, 1965-1975.
12. Giddings, J. C. and Keller, R. A. (Eds.), Chromatographic Science Series, Vols. I-VII, Marcel Dekker, New York, 1965-1977.
13. Leathard, D. A. and Shurlock, B. C., Identification Techniques in Gas Chromatography, Wiley, London, 1970.
14. Gudzinowicz, B. J., Gas Chromatographic Analysis of Drugs and Pesticides, Marcel Dekker, New York, 1967.

B. Mass Spectrometry

1. Roboz, J., Introduction to Mass Spectrometry: Instrumentation and Techniques, Wiley, New York, 1968.
2. Hamming, M. C. and Foster, N. G., Interpretation of Mass Spectra of Organic Compounds, Academic Press, New York, 1972.
3. Gudzinowicz, B. J., Gudzinowicz, M. J., and Martin, H. F., Fundamentals of Integrated GC-MS, Part II, Marcel Dekker, New York, 1976.
4. Biemann, K., Mass Spectrometry: Organic Chemistry Applications, McGraw-Hill, New York, 1962.
5. Beynon, J. H., Mass Spectrometry and Its Applications to Organic Chemistry, Elsevier, Amsterdam, 1960.
6. Reed, R. I., Applications of Mass Spectrometry to Organic Chemistry, Academic Press, New York, 1966.
7. McLafferty, F. W., Interpretation of Mass Spectra, W. A. Benjamin, New York, 1966.
8. Waller, G. R. (Ed.), Biochemical Application of Mass Spectrometry, Wiley, New York, 1972.
9. Johnstone, R. A. W., Mass Spectrometry for Organic Chemists, Cambridge University Press, New York, 1972.
10. Budzikiewicz, H., Djerassi, C., and Williams, D. H., Interpretation of Mass Spectra of Organic Compounds, Holden-Day, San Francisco, 1964.
11. Budzikiewicz, H., Djerassi, C., and Williams, D. H., Mass Spectrometry of Organic Compounds, Holden-Day, San Francisco, 1967.

12. Field, F. H. and Franklin, J. L., Electron Impact Phenomena, Academic Press, New York, 1957.
13. Kiser, R. W., Introduction to Mass Spectrometry and Its Applications, Prentice-Hall, Englewood Cliffs, N.J., 1965.
14. Beynon, J. H., Saunders, R. A., and Williams, A. E., The Mass Spectra of Organic Molecules, Elsevier, New York, 1968.
15. Silverstein, R. M. and Bassler, G. C., Spectrometric Identification of Organic Compounds, Wiley, New York, 1963.
16. Beckey, H. D., Field Ionization Mass Spectrometry, Pergamon Press, Braunschweig, 1971.
17. Budzikiewicz, H., Djerassi, C., and Williams, D. H., Structure Elucidation of Natural Products by Mass Spectrometry, Vols. I–II, Holden-Day, San Francisco, 1964.
18. Hill, H. C., Introduction to Mass Spectrometry, Heyden, London, 1966.
19. Spiteller, G., Massenspektrometrische Strukuturanalyse Organischer Verbindungen, Verlag Chemie, Weinheim, 1966.
20. Creswell, C. J., Runquist, O., and Campbell, M. M., Spectral Analysis of Organic Compounds: An Introductory Programmed Text, 2nd ed., Burgess Publ. Co., Minneapolis, Minn., 1972.
21. Barnard, G. P., Modern Mass Spectrometry, Institute of Physics, London, 1953.

C. Integrated GC-MS

1. McFadden, W., Techniques of Combined Gas Chromatography/Mass Spectrometry, Wiley, New York, 1973.
2. Gudzinowicz, B. J., Gudzinowicz, M. J., and Martin, H. F., Fundamentals of Integrated GC-MS, Part III, Marcel Dekker, New York, 1977.
3. Merritt, C., Jr., Appl. Spectrosc. Rev., 3, 263 (1970).
4. McFadden, W. H., Separation Sci., 1, 723 (1966).
5. Watson, J. T., in L. S. Ettre and W. H. McFadden (Eds.), Ancillary Techniques of Gas Chromatography, Wiley, New York, 1969.
6. Ryhage, R., Quart. Rev. Biophys., 6, 311 (1973).
7. Karasek, F. W., Anal. Chem., 44, 32A (1972).
8. Fenselau, C., Appl. Spectrosc., 28, 305 (1974).

D. Pharmacology/Pharmacokinetics

1. Notari, R. E., Biopharmaceutics and Pharmacokinetics: An Introduction, 2nd ed., Marcel Dekker, New York, 1975.

2. Gibaldi, M. and Perrier, D., Pharmacokinetics, Marcel Dekker, New York, 1975.
3. Goldstein, A., Aronow, L., and Kalman, S. M., Principles of Drug Action: The Basis of Pharmacology, 2nd ed., Wiley, New York, 1974.
4. Goodman, L. S. and Gilman, A. (Eds.), The Pharmacological Basis of Therapeutics, 4th ed., Macmillan, New York, 1970.
5. Wilson, C. O., Gisvold, O., and Doerge, R. F. (Eds.), Textbook of Organic Medicinal and Pharmaceutical Chemistry, 6th ed., Lippincott, Philadelphia, 1971.
6. Levine, R. R., Pharmacology: Drug Actions and Reactions, Little, Brown, Boston, 1973.
7. Melmon, K. L. and Morrelli, H. F. (Eds.), Clinical Pharmacology: Basic Principles in Therapeutics, Macmillan, New York, 1972.
8. La Du, B. N., Mandel, H. G., and Way, E. L. (Eds.), Fundamentals of Drug Metabolism and Drug Disposition, Williams & Wilkins, Baltimore, 1971.
9. Goth, A., Medical Pharmacology: Principles and Concepts, 7th ed., C. V. Mosby, St. Louis, 1974.
10. Meyers, F. H., Jawetz, E., and Goldfien, A., Review of Medical Pharmacology, 4th ed., Lange, Los Altos, Calif., 1974.
11. Garattini, S. and Shore, P. A. (Eds.), Advances in Pharmacology and Chemotherapy, Vols. I-XI, Academic Press, New York, 1962-1973.
12. Ellis, G. P. and West, W. B. (Eds.), Progress in Medicinal Chemistry, Vols. I-IX, Butterworths, London, 1961-1973.

Chapter 2

ETHYL ALCOHOL AND VOLATILE TRACE COMPONENTS IN BREATH, BODY FLUIDS, AND BODY TISSUES

I. INTRODUCTION

Classified as a contemporary drug of abuse, ethyl alcohol (a primary and continuous depressant of the central nervous system) impairs motor coordination and decreases mental acuity even if ingested in small amounts while concurrently inducing a state of euphoria and releasing one from inhibitory attitudes. As noted by Goth [1], "it is often forgotten that alcoholism is still the most serious form of drug abuse in western society. It is estimated that over 9 million Americans are alcoholics, and that over 50% of crimes and over 50% of highway accidents in the United States are alcohol-related." In terms of addiction, ethyl alcohol is of far greater social and medical importance than all other drugs combined [2]. Owing to the widespread, increasing use of other pharmaceutical agents (barbiturates, tranquilizers, anesthetics, oral hypoglycemic drugs, anticonvulsants, anticoagulants) [3], the interaction of drugs is being more fully recognized today as a hazard [4] contributing to automobile fatalities, accidental and suicidal deaths [5], etc. Furthermore, the concurrent use of alcohol with other drugs may alter the therapeutic effect of an agent. Additive effects are seen with excessive and sometimes ordinary doses of central nervous system depressants (i.e., narcotics, sedatives, antianxiety drugs, tranquilizers, and antihistamines). Because of the problems associated with polypharmacy, gas chromatography as well as the integrated GC-MS system are techniques that are extremely suited for monitoring drug abuse as well as determining the individual drug components in mixtures isolated from biological fluids.

On the other hand, the methodology can be adapted to determine trace constituents at parts-per-billion levels in human body fluids such as blood, urine, cerebrospinal fluid, and expired air (breath). As pointed out by Novotny et al. [6], modern chromatography possesses unique potential because, owing to the extreme complexity of the defects in fundamental biochemical processes in living cells that result in metabolic disorders, it provides the means by which several hundred constituents can be separated, detected, and quantitated from one sample. Such analytical data, generally referred to as "metabolic profiles" [7,8], show great promise for biochemical studies and diagnostic purposes.

II. ETHYL ALCOHOL

Ethyl alcohol is absorbed from the stomach, small intestine, and colon. The rate of absorption of ingested alcohol is influenced by two factors: (1) the presence of food in the stomach and (2) the ingested alcohol's concentration. After absorption, it is rather uniformly distributed throughout all body fluids and tissues. If the alcohol level in blood is assigned a value of 1.00, its relative concentrations in various body fluids, tissues, and alveolar air are [1]: urine, 1.30; serum, 1.15; spinal fluid, 1.15; saliva, 1.30; kidney, 0.83; brain or liver, 0.85 to 0.90; alveolar air, 0.00048. In turn, the blood level achieved depends on several factors: (1) the amount of alcohol ingested, (2) the rate of ingestion, (3) the rate of absorption, (4) the percentage of total body water and body weight, (5) the rate of distribution, and (6) the rate of metabolism.

Of the total alcohol that enters the body, about 90 to 98% is completely oxidized, the end products being carbon dioxide and water. The unaltered alcohol (2 to 10%) is normally exhaled from the lungs, excreted in the urine, or lost via perspiration. Its rate of oxidation is a linear function of time, whereas the amount of alcohol oxidized per unit of time is nearly proportional to body weight and probably to liver weight [9]. Although initial alcohol oxidation occurs chiefly in the liver to yield ATP equivalent to approximately 7 kcal/g of ethanol, the first oxidative step as shown in Figure 2.1 is its conversion to acetaldehyde by alcohol dehydrogenase, which is a zinc-containing enzyme that utilizes NAD as the hydrogen acceptor and is

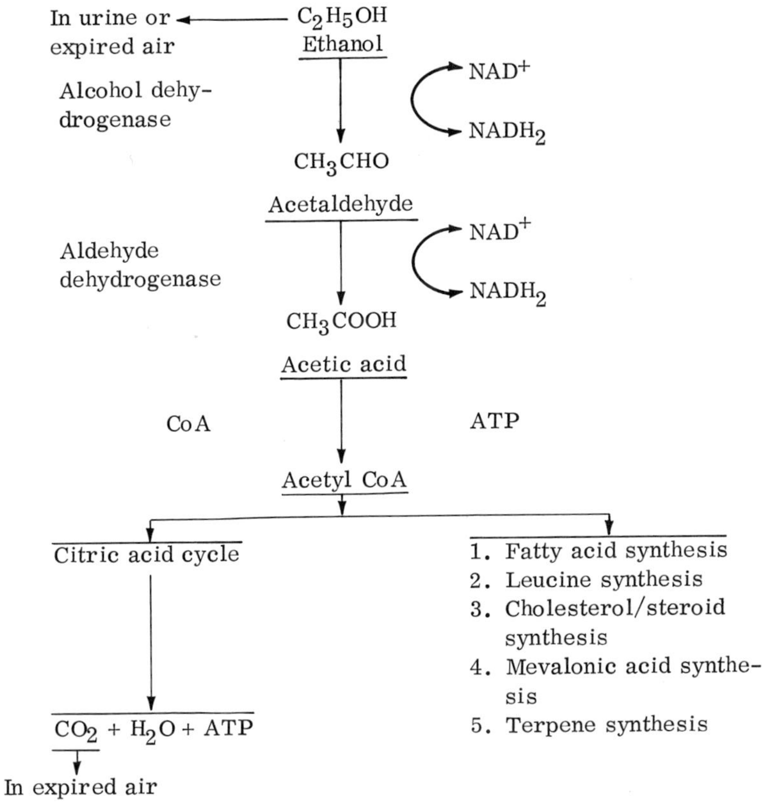

Figure 2.1. The metabolism of ethanol.

dependent on free sulfhydryl groups in the protein for activity [10]. The acetaldehyde is converted, either via a direct pathway or acetic acid, to acetyl coenzyme A (CoA), whose fate is predictable, namely, oxidation through the citric acid cycle or participation in various anobolic reactions in the synthesis of fatty acids, cholesterol, and other tissue constituents.

On the other hand, methanol is metabolized by hepatic catalase to CO_2 at a rate which is only one-seventh that for ethanol. The primary metabolites from methanol are formaldehyde and formic acid, as indicated in Figure 2.2; an aqueous mixture of MeOH and HCHO is easily separated on 60-80 mesh Carbosieve-B (Fig. 2.3).

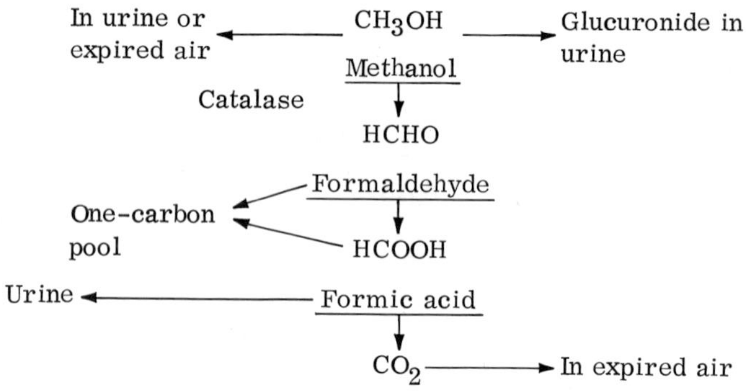

Figure 2.2. The metabolism of methanol.

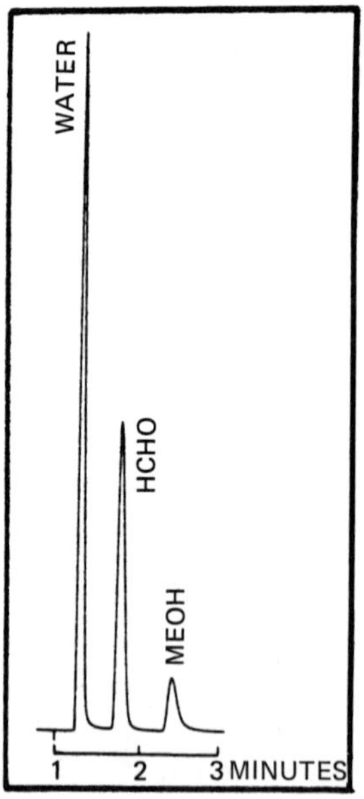

Figure 2.3. Separation of water, methanol, and formaldehyde. Column: 4 ft x 3 mm glass with Carbosieve-B 60/80 mesh; col. temp.: 200°C; flow rate: 75 ml/min helium; det: thermal conductivity cell, 200 MA; sample size: 0.5 μl; (Fisher-Victoreen Gas Chromatograph Model 4400). Courtesy of Fisher Scientific Co.

The metabolic pathway for normal ethanol oxidation can be altered chemically by the administration of disulfiram (Antabuse), calcium carbimide (Temposil), metronidazole (Flagyl) or, rarely, the antidiabetic sulfonylureas such as tolbutamide (Orinase).

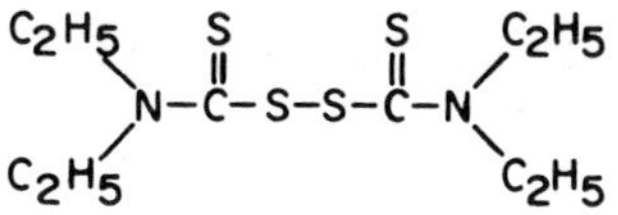

Disulfiram

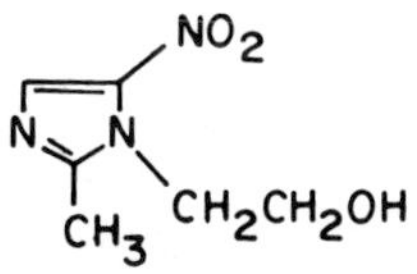

Metronidazole

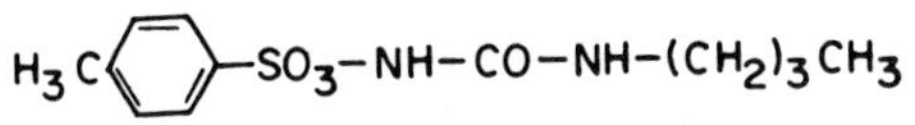

Tolbutamide

Temposil contains citric acid in two parts by weight to one part of calcium cyanamide suitably purified for drug use.

Disulfiram promotes the accumulation of acetaldehyde in the blood, apparently inhibiting aldehyde dehydrogenase in the second step of ethanol metabolism.

Rubin and Lieber [2] tested the effects of chronic administration of ethanol on hepatic smooth endoplasmic reticulum (SER) to see if they were comparable to those of a drug such as phenobarbital or, as expressed by the authors, they "embarked on a series of experiments to determine whether, in addition to effects on the central nervous system, the influence of acute and chronic ethanol ingestion on the response to drugs might be explained by an interaction of ethanol and hepatic microsomes. In other words, we posed the question of whether ethanol could be considered to be a drug in the same way that phenobarbital is."

From their work, which confirmed data reported by other scientists [11-32] and previous studies performed by Rubin, Lieber, and co-workers [33-49], they concluded that ethanol should be included in the list of drugs that are metabolized by hepatic microsomes. Furthermore, chronic administration of ethanol and phenobarbital both produced the following:

1. Increased smooth endoplasmic reticulum
2. Increased activities of drug-metabolizing enzymes in the hepatic microsomes, as illustrated in Figure 2.4.

3. Acceleration of its own metabolism
4. Increased rate of metabolism of other drugs
5. Increase in cholesterol biosynthesis [32,49]

On the other hand, acute administration of ethanol or other compounds leads to inhibition of the metabolism of other drugs by hepatic microsomes and to interference with in vivo drug metabolism as shown in Figures 2.5 and 2.6. Furthermore, they noted that both ethanol and other drugs bind to microsomal P450; simultaneous addition of certain drugs and ethanol to hepatic microsomes leads to mutual inhibition of microsomal binding.

As noted above, concurrent use of alcohol with other drugs may alter the effect of the therapeutic agent. Additive effects are seen with excessive and sometimes ordinary doses of central nervous systems depressants (e.g., narcotics, sedatives, antianxiety drugs, tranquilizers, and antihistamines). The interactions vary from one individual to another, with the amount of alcohol consumed, and with the duration of use of large amounts of alcohol. With intoxicating amounts, hepatic metabolism of many drugs may be inhibited, and an increased therapeutic or toxic effect of the drug may be seen. However, in some individuals who drink more than 8 oz of 80-proof whiskey or the equivalent daily for a prolonged period, hepatic metabolism of drugs may be increased. This will diminish the therapeutic effects of many drugs. Also, alcoholic beverages can vary widely, and

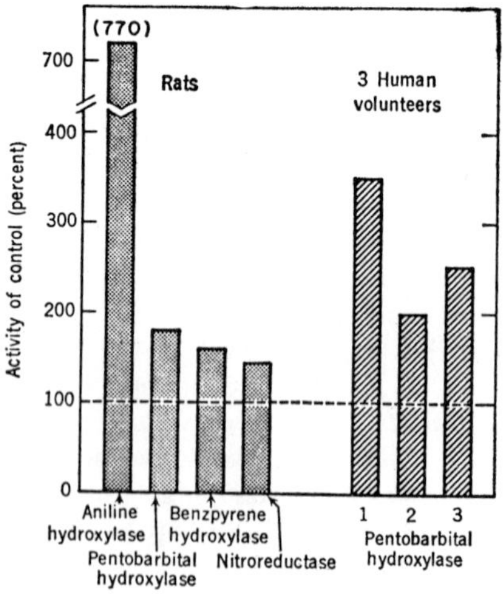

Figure 2.4. Effect of chronic administration of ethanol on activities of hepatic drug-metabolizing enzymes in rats and human volunteers. The activity of every enzyme studied was increased by chronic ingestion of ethanol. From Rubin and Lieber [2], courtesy of Science (A.A.A.S.).

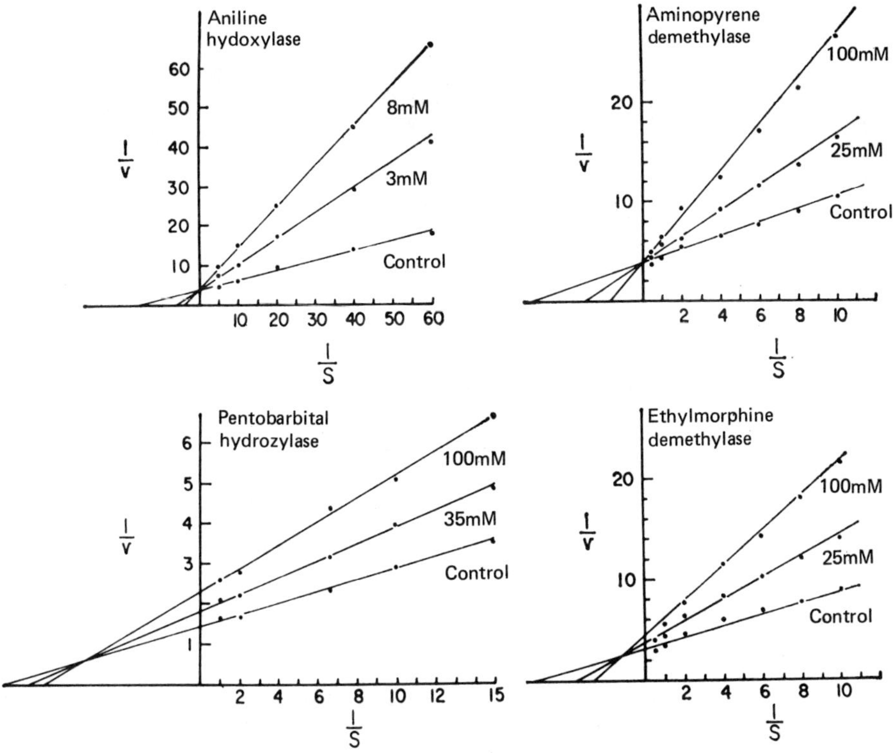

Figure 2.5. Lineweaver-Burk plots showing inhibition of various hepatic drug-metabolizing enzymes by ethanol. Intersection of the lines on the ordinate, in the cases of aniline hydroxylase and aminopyrene demethylase, indicate competitive inhibition. Intersection of the lines between the ordinate and the abscissa, in the cases of pentobarbital hydroxylase and ethylmorphine demethylase, indicate a mixed type of inhibition. The concentrations of ethanol are shown in each graph. (S indicates millimolar concentration of substrate; V indicates velocity of the reaction.) From Rubin et al. [42], courtesy of American Journal of Medicine.

this fact should be taken into consideration when assessing such potential interactions with alcohol. Wines possess a relatively high buffer capacity and hence are absorbed less rapidly. In Table 2.1 are listed some well-documented interactions of drugs with alcohol compiled by the Rhode Island Hospital's Department of Pharmacy.

Several investigators [50] have suggested that alcohol dependence and tolerance are etiologically related to morphine tolerance and dependence, based on structural resemblance between THP and morphine. Holtz, Stock, and Westermann [51] noted that, via a Pictet-Spengler condensa-

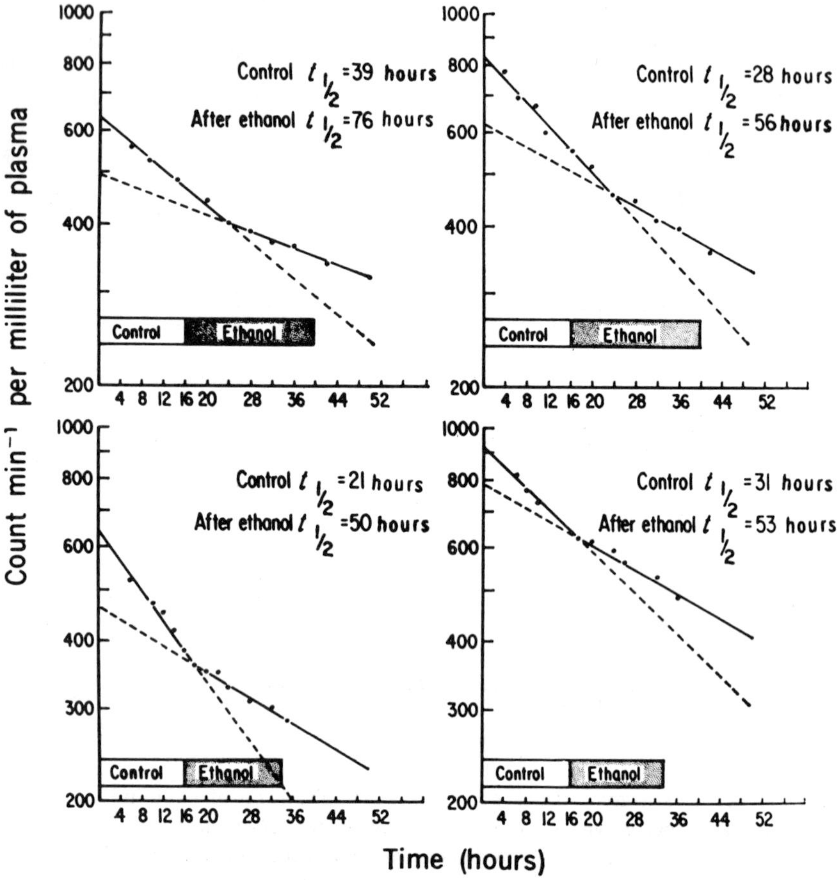

Figure 2.6. Effect of acute ethanol intoxication on the disappearance of pentobarbital from the blood of four human volunteers. After ethanol administration (1 g/kg, followed by 30 g every 2 hr), the half-life of pentobarbital in the blood was approximately doubled. Solid lines are plotted from data points. Dashed lines are extrapolated. From Rubin et al. [42], courtesy of American Journal of Medicine.

tion, an alkaloid, tetrahydropapaveroline (THP), could be formed from dopamine and its corresponding aldehyde. In the presence of alcohol, acetaldehyde competitively inhibits aldehyde dehydrogenase, thus favoring THP formation [52]. However, when alcohol is ingested, it is also possible to form the alkaloid salsolinol (the condensation product of acetaldehyde and dopamine). Davis and Walsh [50] have shown that THP in rats is considerably metabolized to opium alkaloids and have reported cross-tolerance between morphine and alcohol in mice. This conclusion is doubtful, since Goldstein and Judson [53] showed that animals dependent on opiates yielded a well-defined withdrawal jumping syndrome when given naloxone, but the alcohol-dependent mice showed no naloxone-induced opiate withdrawal signs. From this, there appears to be no link between opiate dependence and alcohol dependence.

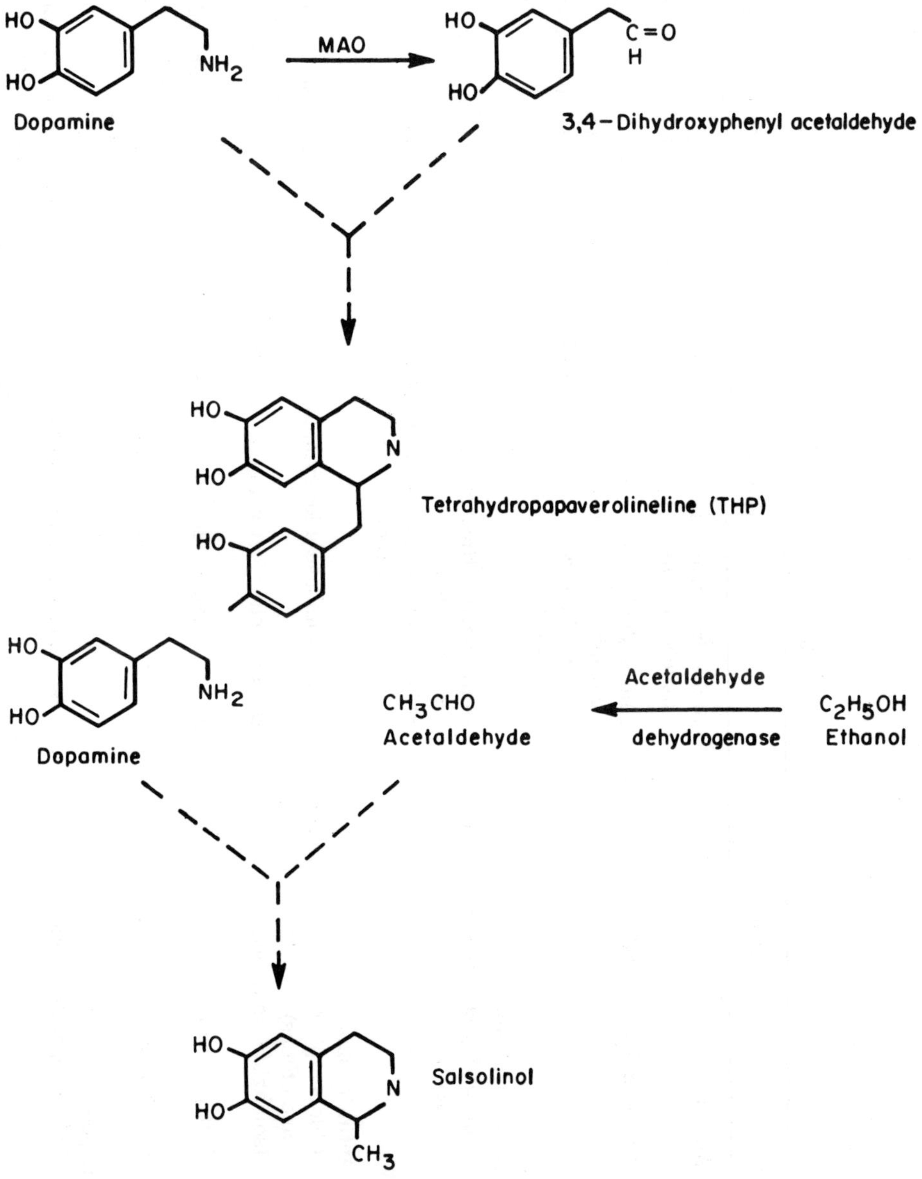

TABLE 2.1

Drug Interactions with Alcohol[a]

Drug	Effect	Probable Mechanism
Antabuse	Abdominal cramps, flushing, vomiting, psychotic episodes, confusion	Inhibits intermediary metabolism of alcohol
Anticoagulants, oral	Diminished anticoagulant effect with chronic alcohol abuse	Enhanced microsomal enzyme activity
	Enhanced anticoagulant effect with acute intoxication	Reduced metabolism
Anticonvulsants		
Diphenylhydantoin (Dilantin and others)	Diminished anticonvulsant effect with chronic alcohol abuse	Enhanced microsomal enzyme activity
	Enhanced anticonvulsant effect with acute intoxication	Reduced metabolism
Antimicrobials		
Chloramphenicol (Chloromycetin and others)	Minor Antabuse-like symptoms	Inhibits intermediary metabolism of alcohol
Isoniazid (many brands)	Diminished effect with chronic abuse	Enhanced microsomal enzyme activity
Metronidazole (Flagyl)	Minor Antabuse-like symptoms	Inhibits intermediary metabolism of alcohol
Quinacrine (Atabrine)	Minor Antabuse-like symptoms	Inhibits intermediary metabolism of alcohol
Hypoglycemics		
Chlorpropamide (Diabinese)	Minor Anabuse-like symptoms	Inhibits intermediary metabolism of alcohol

Phenformin (DBI; Meltrol)	Lactic acidosis	Synergy
Tolbutamide (Orinase)	Diminished hypoglycemic effect with chronic alcohol abuse	Enhanced microsomal enzyme activity
	Enhanced hypoglycemic effect with ingestion of alcohol, particularly in fasting patients	Suppression of gluconeogensis
	Minor Antabuse-like symptoms	Inhibits intermediary metabolism of alcohol
MAO inhibitors	Hypertensive crisis	Inhibit metabolism of tyramine in Chianti wine and some other alcoholic beverages
Salicylates	Gastrointestinal bleeding	Additive
Sedatives and tranquilizers		
Barbiturates	Diminished sedative effect with chronic alcohol abuse	Enhanced microsomal enzyme activity
	Enhanced central nervous system depression with acute intoxication	Additive; reduced metabolism
Chloral hydrate (Noctec and others)	Prolonged hypnotic effect	Mutual potentiation
Diazepam (Valium)	Enhanced central nervous system depression	Additive
Meprobamate (Miltown and others)	Diminished sedative effect with chronic alcohol abuse	Enhanced microsomal enzyme activity
	Enhanced central nervous system depression with acute intoxication	Additive; reduced metabolism

[a]Courtesy of Department of Pharmacy, Rhode Island Hospital, Providence, R.I.

As late as May 1972, the controversy over the significance of morphine-like alkaloids produced by condensation with acetaldehyde seemed no nearer resolution after a symposium on isoquinolines and other morphine-like substances. Seixas [54] offered the following as a "fair summation" of the status of the subject:

> Few investigators feel that these putative morphine-like alkaloids correlate in any convincing sense with the clinical symptoms of alcoholism. What the experiments mean still needs a lot of working out. Nevertheless, just the facts that these compounds can be made to exist--even if, so far, only under somewhat unphysiological in vitro conditions--and that they've been found for each of the biogenic amines provide an exciting realm for chemical speculation, one that must certainly be explored.

Ahtee [55] reported data linking the biogenic amine serotonin (5-hydroxytryptamine) with alcoholism based on brain concentration studies of serotonin (5-HT) and 5-hydroxyindolylacetic acid (5-HIAA, a serotonin metabolite) in two strains of rats that had been selected for 16 to 20 generations for their preference for ethanol or water. Results showed that brain concentrations of 5-HT and 5-HIAA were 15 to 20% higher in ethanol-preferring rats than in water-preferring rats, even when rats of both strains were "naive" with respect to alcohol. If given a free choice between water and ethanol (10% v/v) for a month, Ahtee noted that 5-HT brain levels were 31% higher in addicted than in nonaddicted rats, whereas the 5-HIAA level was elevated by only 10%. Such was not the case for nonalcoholic rats. They showed no significant changes in 5-HT and 5-HIAA concentrations. These preliminary findings support the idea that serotonin is involved in the mechanism of alcohol preference, lending credibility as well to the concept that susceptibility to alcohol addiction can be greatly influenced by genetic factors.

Many alcoholics suffer from disturbances of memory orientation, judgment, comprehension, and effect, and recent research by Tewari and Noble [56] has shown a strong involvement of nucleic acid and protein synthesis in memory function, learning, and behavior. Since the recuperative powers of the adult brain are very limited, it is possible that long-term use of alcohol can cause enough damage to the brain to impair significantly, perhaps permanently, the behavior of these subjects.

Using principles of chemical kinetics related to the absorption and utilization or elimination of drugs, Calder [57] illustrated these very effectively by the time evolution of ethanol. Whereas the absorption of ethanol through the stomach and intestinal wall is a rapid first-order process [58], its rate of elimination from the body is independent of its concentration, the elimination being a zero-order process.

Calder depicted the overall mechanism for the metabolism of ethanol as

$$G \xrightarrow{k_{in}} B \text{ (first order)} \tag{2.1}$$

$$B \xrightarrow[k_{out}]{} E \text{ (zero-order)} \tag{2.2}$$

and

$$\frac{d(G)}{dt} = -k_{in}(G) \tag{2.3}$$

$$\frac{d(B)}{dt} = k_{in}(G) - k_{out} \tag{2.4}$$

$$\frac{d(E)}{dt} = k_{out} \tag{2.5}$$

Integrating for the ethanol concentration in the blood (B), assuming that (B) = 0 at t = 0, one obtains

$$(B) = (G)_0 \, [1 - \exp(-k_{in}t)] - k_{out}t \tag{2.6}$$

Since d(B)/dt = 0 implies the condition of maximum ethanol blood concentration, this can be also expressed as

$$(G)_0 = (\frac{k_{out}}{k_{in}}) \exp(k_{in}t_{max}) \tag{2.7}$$

or

$$(B)_{max} = (G)_0 - (\frac{k_{out}}{k_{in}}) - (k_{out}t_{max}) \cong (G)_0 \tag{2.8}$$

After a single rapid ingestion of alcohol, the time dependence of its blood concentration is an exponential increase for $t < t_{max}$ with a linear decrease for $t > t_{max}$.

If one assumes the following conversion factors (Table 2.2) and values of constants ($k_{in} = 10\ hr^{-1}$, $k_{out} = 0.192$ g of ethanol per liter of body fluid per hour) to be used in the analysis of the time evolution of ethanol in the body, one can plot the time evolution of ethanol in the blood as shown in Figure 2.7. In Figure 2.7, one notes a short induction period for ethanol before minimum effective concentration is achieved; this effective concentration of 1 g/liter corresponds to the legal definition of intoxication.

TABLE 2.2

Conversion Factors and Values of Constants Used in the Analysis of the Time Evolution of Ethanol in the Body[a,b]

Ethanol (g)	Concentration of ethanol in body fluid (g/liter)	Equivalent volume of 80-proof (40% v/v ethanol) liquor: ml	oz	Qualitative physiological and psychological response
20	0.5	63	2.1	Sense of euphoria
40	1.0	127	4.3	Mild to medium disturbance in motor coordination; legally intoxicated.
60	1.5	190	6.4	Overt lack of coordination and slurring of speech
80	2.0	254	8.6	Amnesia
100	2.5	317	10.7	Induced sleep or unconsciousness

[a]From Calder [57], courtesy of Journal of Chemical Education.
[b]$k_{in} = 10\ hr^{-1}$; $k_{out} = 0.192$ g of ethanol per liter of body fluid per hour = 10 ml ethanol/hr, volume of body fluid = 40 liters.

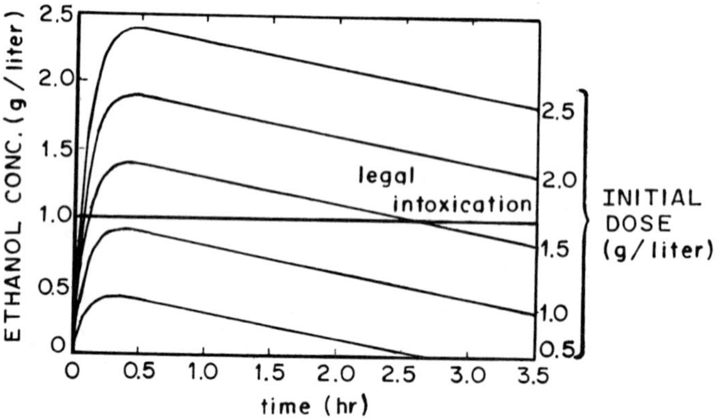

Figure 2.7. The time evolution of ethanol in the body, assuming instantaneous ingestion. From Calder [57], courtesy of Journal of Chemical Education.

As a useful reference point, this induction time t_{ind} is given by

$$t_{ind} \cong k_{in}^{-1} \ln(\frac{G_0}{G_0 - 1}) \qquad (2.9)$$

where G_0 is the initial dose of ethanol expressed in grams per liter of body fluid.

As pointed out by Calder, although it is not a sensitive function of dosage, the time at which maximum blood concentration is reached may be expressed as

$$t_{max} = k_{in}^{-1} \ln(\frac{G_0 k_{in}}{k_{out}}) \qquad (2.10)$$

The maximum blood concentration is essentially equal to the quantity of ethanol ingested, which is dictated by its slow rate of elimination.

Associated with the elimination process are two characteristic times: the sobriety time t_{sob}, which is the time needed for the alcohol level to decrease below the legal intoxication level (1 g/liter) after an initial dose in excess of that amount is ingested; and the elimination time t_{el}, which is the time required to eliminate the ethanol completely from the body.

$$t_{sob} \cong (G_0 - 1)k_{out} = 1.22(OZ - 4.3) \qquad (2.11)$$

where G_0 is expressed in grams per liter and OZ is ounces of 80-proof liquor consumed. Based on this equation, Calder calculated that for every ounce of 80-proof liquor ingested in excess of the 4.3 oz (Table 2.2), it would require 1 hr and 15 min to become legally sober again, or expressed in another way, only a small additional dosage is necessary to maintain the blood concentration at the same level (that is, 10 ml of ethanol per hour = 0.85 oz of 80-proof alcohol per hour) once the effective concentration is reached, this being only 20% of the initial dosage.

The elimination time or residence time of ethanol in the body, which essentially parallels the recuperation or hangover time, is given by

$$t_{el} = G_0 k_{out}^{-1} = OZ \times 1.22 \qquad (2.12)$$

A more realistic example is that of an individual at a cocktail party whose consumption of ethanol occurs at a constant rate, after which consumption ceases. For this model, Calder proposed a more suitable mechanism,

$$C \xrightarrow{k_{dose}} G \text{ (variable zero order)} \qquad (2.13)$$

$$G \xrightarrow{k_{in}} B \text{ (rapid first order)} \qquad (2.14)$$

$$B \xrightarrow{k_{out}} E \text{ (slow zero order)} \tag{2.15}$$

where (G) is again the alcohol concentration ingested and (B) is the blood concentration.

Solving for (B) using appropriate differential equations for this mechanism, one arrives at the following expression:

$$(B) - (k_{dose} = k_{out})t - \frac{k_{dose}}{k_{in}}[1 - \exp(-k_{in}t)] \tag{2.16}$$

for $t < t_{max}$.

Because the ethanol transport rate across the gastrointestinal membrane is rapid relative to the dose rate during the ingestion period, Eq. (2.16) may be approximated by

$$(B) \cong (k_{dose} - k_{out})t \tag{2.17}$$

for $t < t_{max}$.

When ingestion is terminated (that is, for $t > t_{max}$),

$$(B) = (B)_{max} - k_{out}t \tag{2.18}$$

As illustrated by Calder, this is shown in Figure 2.8 for two different ingestion rates, 1 g/liter/hr and 1.5 g/liter/hr for a period of 1 hr, and compared with the corresponding blood alcohol concentration curves for the instantaneous ingestion.

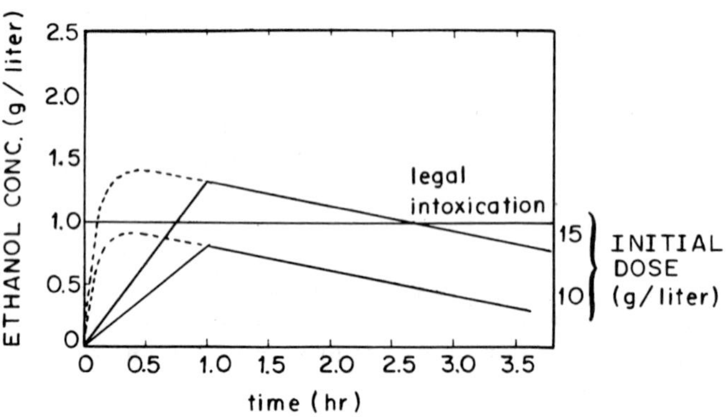

Figure 2.8. The time evolution of ethanol in the body assuming a constant rate of ingestion for 1 hr and subsequent termination. From Calder [57], courtesy of Journal of Chemical Education.

From this mathematical treatment, Calder made the following conclusions.

1. Ethanol maximum blood concentration is not significantly altered or decreased by a prolonged ingestion period unless it is of the order of t_{sob}. Hence, the patient drinker who "nurses" his alcohol for 1 hr and the compulsive drinker will both become equally intoxicated if the physiological response and the alcohol intake by each person is the same.

2. By prolonged ingestion, the length of time that the blood alcohol concentration remains above the effective level (1 g/liter) is significantly reduced; this shortening of the intoxication period occurs in the early stages with no reduction of t_{sob} resulting, since this is essentially governed by the invariant metabolic rate. Although the patient drinker will become equally intoxicated more slowly, the time to regain sobriety will be the same as that for the compulsive drinker.

A. Chromatographic Methods of Analysis for Ethanol in Various Specimens

Since 1958, when Cadman and Johns [59] proposed a method for the separation and quantitative analysis of ethanol as well as related volatiles using gas chromatography, the gas chromatographic measurement of ethanol in various biological media (blood, breath, tissue, urine) has been essentially performed by one of four major methods: (1) extraction, (2) distillation, (3) direct injection, or (4) headspace. The trend toward automation has recently been reviewed in great detail by Cravey, Reed, and Sedgwick [60], Jain and Cravey [61-63], Patterson [64], and Dubowski [65]; the automated alcohol systems resulting from the concentrated effort of law enforcement, precipitated by public demand, to control the drinking driver who has imposed monumental work loads on the forensic laboratories analyzing these specimens. Furthermore, as emphasized by Jain and Cravey [61], it follows that precision, specificity, and accuracy are analytical criteria that are demanded of these automated techniques, especially if the results are to be used for legal purposes.

1. Extraction Methods [59,60,66-74,82]

In 1958, Cadman and Johns [59] developed a gas chromatographic procedure for determining the ethanol content in blood as well as related volatile compounds using a Beckman GC-2 gas chromatograph equipped and operated with the following accessories and parameters:

1. Column dimensions and packing: 366-cm x 0.63-cm o.d. SS tubing filled with a 28% (w/w) mixture of liquid stationary phases (Flexol

8N8, diisodecylphthalate, and polyethylene glycol PEG 600 mixed in a 15:10:3 ratio by weight) coated on 42-60 mesh C-22 firebrick.

2. Operational temperatures: (a) injection port, 160°C; (b) column and detector: 110°C.
3. Carrier gas and flow rate: helium at 60 ml/min.
4. Detector: four-filament hot-wire (tungsten) thermal conductivity cell at 330 mA.
5. Sample size: 35 μl.

The extraction procedure is rather simple and straightforward. It consists of pipetting 1 ml of blood into a 3-ml centrifuge tube (screw-capped) containing 1 ml of n-propyl acetate and 1 g of anhydrous potassium carbonate. Capped, mixed, and shaken for 1 min on a mechanical shaker, the specimen is centrifuged for 2 min to break the emulsion. The n-propylacetate supernatant, which removes about 77% of the alcohol, is decanted into a 1-ml serum bottle. From this extract, a 35-μl aliquot is sampled with a microsyringe and then injected into the gas chromatograph. With the 12-ft column that is required to totally separate ethyl ether, methanol, acetone, ethanol, isopropanol, and n-propylacetate (this being their order of elution), the retention time of ethanol is nearly 8 min, whereas the solvent is completely eluted in 24 min. For quantitative studies, blood alcohol standards are prepared by adding measured amounts of ethanol to blood; these are subsequently extracted and chromatographed to prepare a calibration curve under conditions identical to those used for sample analysis. As noted by Harger and Forney [75], "the authors reported an accuracy of ±3 mg% ethanol, but their results with 16 samples of alcoholic blood showed deviations of 1 to 17 mg% (average 6 mg%) between their gas chromatography results and those by the Kozelka-Hine method."

Lyons and Bard [67] use the same chromatographic equipment described by Cadman and Johns [59], but their column is packed with 15% (w/w) of the mixed liquid stationary phases (same weight ratios maintained) impregnated on 30-60 mesh Chromosorb. For GC analysis, sample preparation consists of the slow addition with shaking of 1 ml of blood to 2 g of a mixture of sodium bisulfite and calcium sulfate. To the above, 0.5 ml of n-butanol is added. The container housing this heterogeneous mixture is then closed, shaken, and centrifuged. Following the transfer of the n-butanol layer to a small vial to which 150 mg of anhydrous potassium carbonate is added, 5 μl of this solution is injected into the chromatograph after the mixture is allowed to stand for 5 min so that the drying agent can completely remove any water present in the n-butanol phase. Using a calibration curve prepared from known blood alcohol standards subjected to the same extraction-GC analytical procedure, the alcohol concentration in blood is determined.

In 1965, Steinberg et al. [69] modified the procedure of Cadman and Johns by using a gas chromatograph equipped with a flame ionization detector and a 15-ft x 1/8-in. SS column packed with 15% (w/w) Hallcomid M-18 on 60-80 mesh, HMDS-treated Chromosorb W. Davis [68] used the same instrument and column recommended by Cadman and Johns with several exceptions: (1) a hydrogen flame ionization detector was employed rather than the TC detector, (2) the column and injector temperatures were maintained at 105 and 230°C, respectively, and (3) a helium flow rate of 65 ml/min was used.

Laessig [74] demonstrated a submicro sampling and solvent extraction system designed to take advantage of the inherent high specificity and sensitivity of gas chromatography to reduce manipulative steps in the analysis to a minimum; the procedure was applied to the determination of blood alcohol. Generalizing all biomedical blood analyses by gas chromatography as consisting of eight steps, (1) sampling--venipuncture and collection of 1 to 5 ml of blood, (2) measurement--taking an aliquot of blood for analysis, (3) removal of water from sample, (4) solvent extraction of sample constituents into a volatile solvent, (5) addition of an internal standard to the specimen, (6) centrifugation to compact solids and to separate solvent layers, (7) injection of an aliquot into the gas chromatograph, and (8) the analysis and interpretation of the results, Laessig combined steps 1 through 5 into a single operation performed at the time of sampling. The sampling procedure developed was based on the Unopette sampling system of Gerarde [76,77]. For the determination of ethanol in blood, Laessig used 0.7 ml of redistilled n-butanol as a solvent in a Unopette reservoir which contained n-propanol (7.5 μg/ml) as internal standard. Serving two purposes (providing a surface on which precipitated protein and blood cell membranes are adsorbed and a suitable chemical reagent for adsorbing water from the solvent and blood specimen), anhydrous $CaSO_4$ (0.3 ± 0.05 g) is placed into each reservoir. Upon centrifugation, this arrangement permits the drying agent and adsorbed substances to be packed into a firm mass which allows an aliquot to be drawn into an injection syringe (sample size 0.4 μl) without danger of plugging the needle. The criteria for evaluation of internal standard stability and the nonreactivity and stability of the solvent system stored in the Unopettes was the comparison of the ratio of internal standard to solvent peak area over a period of 4 weeks. The mean value of this ratio was 1.25 x 1024, while the standard deviation was 0.06 (5% relative). The results of this study confirmed the compatibility of the solvent-internal standard system with the polyethylene Unopette. A typical gas chromatogram for the analysis of a 0.10% (w/w) ethanol sample is shown in Figure 2.9.

For this investigation, an F&M Model 402 dual-column flame ionization detector, dual-electrometer chromatograph was used. The columns were 6-ft x 4-mm-i.d. U-shaped glass columns packed with 10% (w/w) Igepal

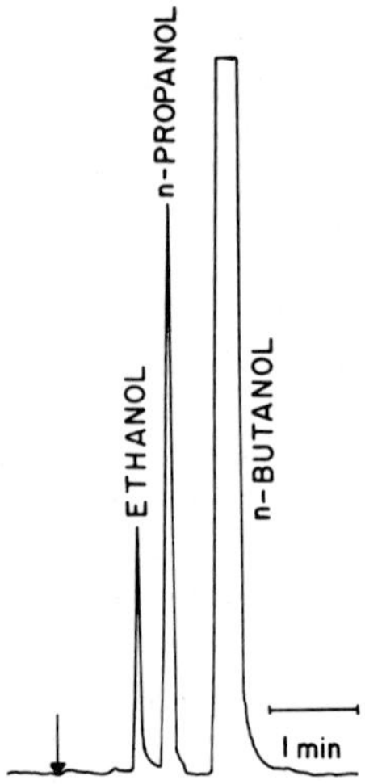

Figure 2.9. Typical chromatographic tracing. 25 μl of blood, 0.7 ml of n-butanol, 0.4 μl injected. Sample 0.100% ethanol (w/w). Sensitivity, ethanol, propanol 10^2 x 1 (with automatic attenuator). Butanol 10^2 x 1024. From Laessig [74], courtesy of Analytical Chemistry.

coated on 80-100 mesh Diatoport S. The operational temperatures used were flash heater, 140°C; detector, 140°C; and column, 110°C. Routinely purged overnight at 110°C and over weekends at 180°C, the column flow rates used for analysis were helium (carrier gas), 60 ml/min; air, 300 ml/min; and hydrogen, 37 ml/min.

The proposed blood ethanol analytical procedure was evaluated with respect to three fundamental criteria: (1) its ability to preserve blood ethanol levels after sampling, until analysis, while samples were stored in the Unopette (demonstrated by the data reported in Table 2.3 on four separate 1-week studies); (2) the accuracy of the method when compared to other accepted techniques for determining blood ethanol; and (3) the precision of the method (see Table 2.4).

The accuracy of the method was compared to the macro gas chromatographic method of Cadman and Johns [59] and the alcohol dehydrogenase procedure of Goldberg and Rydberg [78]. A comparison of the proposed method with the macro GC system yielded a mean absolute error of 0.007% and a mean algebraic error of 0.001%. On the other hand, compared with the enzymatic ADH technique, the mean absolute error was 0.005% and the mean algebraic error was +0.002%. Based on a typical sample value of

TABLE 2.3

Stability of Specimens Stored in Unopettes[a]

Series no.	Mean[b] ratio	Ethanol Calc.[c]	Ethanol Obs.[d]	Error	Rel. error (%)
1	0.696	0.3016	-		
	0.457	0.211	0.198	-0.013	6
	0.261	0.121	0.113	-0.008	6.5
	0	0	0	-	
2	0.581	0.3016	-		
	0.395	0.211	0.205	-0.006	3
	0.249	0.121	0.127	+0.006	5
	0	0	0	-	
3	0.596	0.3016	-		
	0.378	0.211	0.201	-0.010	5
	0.208	0.121	0.117	-0.003	7.5
	0	0	0	-	
4	0.586	0.3016	-		
	0.404	0.211	0.208	-0.003	1.5
	0.242	0.121	0.124	+0.003	2.5
	0	0	0	-	
				Mean	4.0

[a]From Laessig [74], courtesy of Analytical Chemistry.
[b]Ratio ethanol n-propanol peak area.
[c]By dilution, data in weight % ethanol.
[d]By chromatographic analysis, weight % ethanol.

0.136%, the mean absolute errors were 5% (GC) and 4% (ADH), relative--well within the limits normally imposed on chromatographic assays.

From the precision study, Laessig concluded that the method is capable of a routine precision of 3 to 4% (relative), which is comparable to or better than that reported for other methods: 5% [79] and 6% [80] for the macro chromatographic method, 4% [81] for the distillation method, and 5% [81] for the ADH technique.

Roach and Creaven [82] developed a simple, rapid, and accurate GC micro-method for the determination of ethanol and acetaldehyde in blood which permitted the study of ethanol metabolism.

TABLE 2.4

Precision of Samples Taken in Unopette[a, b]

Series no.	Ratio[c] ethanol:propanol mean value	Std. dev.	Number of data
1	0.312	0.009	20
2	0.358	0.008	20
3	0.333	0.014	18
4	0.379	0.008	20
5	0.405	0.009	20
Av. std. dev.		0.010	(3% relative)

[a]From Laessig [74], courtesy of Analytical Chemistry.

[b]Sample from pool of 0.200% ethanol in blood. Experiment repeated five times on three different days. Chromatographic conditions specified in text.

[c]Ratio of peak areas.

Using an F&M model 402 gas chromatograph with a flame ionization detector and a 6-ft x 1/4-in.-i.d., U-shaped glass column packed with pretreated Polypak-2 (80-120 mesh), the helium carrier gas is held at a flow rate of 96 ml/min, with the following operational temperatures maintained: injection port, 110°C; column, 110°C; and detector, 135°C.

Sample preparation is performed in the following manner: Into a 1-ml silicone-rubber stoppered bottle containing 50 μl of internal standard solution (t-butanol, 2.5 mg/ml), are transferred blood samples drawn up into a 50-μl glass disposable microsampling pipette. After this transfer of sample, the bottle is quickly stoppered. Fifty microliters of 5% zinc sulfate and 50 μl of 0.3 N barium hydroxide are injected into the bottle through the stopper; this is followed by centrifugation at 5000 x g for 10 min at 0°C. After centrifugation, 1 μl of the supernatant phase is chromatographed. Under these prescribed GC conditions, the retention times for a mixture containing acetaldehyde, ethanol, acetone, and t-butanol are 2.8, 4.6, 8.9, and 18.7 min, respectively. Recoveries of ethanol and acetaldehyde added to whole blood (ethanol, 760.0 μg/ml; acetaldehyde, 19.9 μg/ml) are 104.2 and 98.5%, respectively; concentrations of acetaldehyde and ethanol remain constant up to 6 hr after sample preparation. As noted by Roach and Creaven, the use of a neutral precipitant eliminates the injection of whole blood onto the column and makes prior equilibration of the sample with the gas phase unnecessary, as is true of the procedure suggested by Truitt and Duritz [83]. This permits samples which remain stable for several hours after preparation to be rapidly and easily prepared.

In their comprehensive gas chromatographic drug screening procedure for the clinical laboratory, Griffiths et al. [72] developed a procedure which is carried out with only one extraction and without derivatization of the substance to be determined. For this investigation, a Bendix model 2500, four-column, four-flame ionization detector, monolinear temperature-programmable gas chromatograph was employed with an Autolab 6300 digital integrator and helium as carrier gas. Three different columns were used for the various drugs listed in Table 2.5: one 6-ft x 2-mm-i.d., U-shaped column packed with Porapak Q; a second with 3% (w/w) OV-17 coated on silanized Chromosorb W (100-120 mesh); and a third with 5% OV-1 (w/w) on silanized 100-120 mesh Chromosorb W. With the use of five standard solutions containing the 19 components determined by the screening program outlined and three internal standard solutions for quantitative investigations, 1 ml each of serum standard and serum sample are diluted with 1 ml of 1 M phosphate buffer (pH - 7.4). Shaking each with 10.0 ml of chloroform, 8.0 ml of chloroform is recovered and split into two equal portions, each of which is evaporated to dryness under a stream of nitrogen at 40 to 50°C in a 2-ml test tube. To one of the dried aliquots is added 12.5 μl of methanolic methyl stearate internal standard solution, whereas the residue in the second aliquot is dissolved with 12.5 μl of methanolic dehydroisoandrosterone internal standard solution.

In a parallel series of standard sample preparations, 10 μl of an alcohol standard or serum are diluted to 50 μl with an aqueous solution containing n-butanol as internal standard while concurrently an additional 10 μl of serum are diluted to 20 μl with aqueous tertiary amyl alcohol internal standard solution. The resulting four solutions are then ready for injection. Thus, with slightly more than 1 ml of blood, a screen for 19 drugs can be performed.

For the first two injections, the column temperature is maintained at 160°C, the injectors at 260°C, and the detectors at 300°C, with a helium carrier gas flow rate of 45 ml/min.

The serum, diluted with internal standard no. 1 (1 mg of dehydroisoandrosterone/ml of MeOH), and alcohol working standard (1 mg each of methanol, ethanol, and isopropyl alcohol/ml of water) are injected on the Porapak Q column; the serum and serum standard extracts which were dissolved in methyl stearate internal standard solution (1 mg of methyl stearate/ml of MeOH) are injected onto the OV-1 column. As noted by Griffiths et al., all components of interest will be eluted from the OV-1 (methyprylon, glutethimide, meprobamate, and the barbiturates except phenobarbital) and Porapak Q (methanol, ethanol, and isopropanol) columns in approximately 20 min for each sample.

TABLE 2.5

Standard Deviation and Coefficient of Variation for Analysis of Each of the Included Drugs[a]

	Drug level (mg/100 ml)	S.D. (mg/100 ml)	N
Methanol	100	6.8	10
Ethanol	100	7.7	10
Isopropanol	100	9.1	10
Methyprylon	0.50	0.057	13
Butabarbital	1.1	0.18	10
Amobarbital	1.0	0.12	10
Pentobarbital	1.1	0.050	-
Secobarbital	1.0	0.19	10
Glutethimide	1.0	0.12	13
Meprobamate	5.0	0.87	10
Chloral hydrate (as trichloroethanol)	11.6	0.74	10
Ethchlorvynol	15.0	0.67	10
Methaqualone	0.6	0.024	10
Oxazepam	2.4	0.32	10
Diazepam	1.6	0.12	11
Chlordiazepoxide	2.0	0.07	10
Flurazepam	2.4	0.19	10
Phenobarbital[b]	2.0	0.22	12
Diphenylhydantoin[b]	2.0	0.22	12

[a]From Griffiths et al. [72], courtesy of Clinical Biochemistry.

[b]Extracted by procedure described above, and analyzed by the method of Barret [84].

By raising the oven (hence, the column) temperature to 210°C, the serum and standard (both diluted with aqueous tertiary amyl alcohol solution standard containing 0.1 mg of tertiary amyl alcohol/ml of water) are injected into the Porapak Q column. The serum and serum standard extracts, which have been taken up in the internal standard dehydroisoandrosterone solution (internal standard no. 1) are chromatographed on the OV-17 column. All injections are 1 to 4 μl. Within 40 min, trichloroethanol (serum component of chloral hydrate) and ethchlorvynol are easily eluted from Porapak Q, while the separation of phenobarbital, methaqualone, oxazepam, diphenylhydantoin, diazepam, chlordiazepoxide, and flurazepam is carried out on the OV-17 column.

Recovery studies of drugs added to drug-free serum and carried through the analytical procedure versus standard solutions prepared directly are shown in Table 2.6; the standard deviation and coefficient of variation for each of the drugs is shown in Table 2.5.

TABLE 2.6

Recovery of Drugs Added to Drug-free Serum and Carried Through the Analytical Procedure versus Standard Solution Prepared Directly[a]

Methanol	100%	Chloral hydrate (as trichloroethanol)	101%
Ethanol	100%	Ethchlorvynol	102%
Isopropanol	103%	Methaqualone	83%
Methyprylon	88%	Oxazepam	165%
Butabarbital	96%	Diazepam	80%
Amobarbital	109%	Chlordiazepoxide	132%
Pentobarbital	102%	Flurazepam	84%
Secobarbital	105%	Phenobarbital[b]	83%
Glutethimide	77%	Diphenylhydantoin[b]	83%
Meprobamate	73%		

[a]From Griffiths et al. [72], courtesy of Clinical Biochemistry.
[b]Extracted by procedure described above, and analyzed by the method of Barrett [84].

The elution order of the drugs on the various columns is as follows:

1. Porapak Q at 160°C; methanol, ethanol, isopropanol, and internal standard (n-butanol)
2. Porapak Q at 210°C: internal standard (tertiary amyl alcohol), trichloroethanol, and ethchlorvynol
3. OV-1 at 160°C: methypyrlon, butabarbital, amobarbital, pentobarbital, secobarbital, meprobamate, glutethimide, and internal standard (methyl stearate)
4. OV-17 at 210°C: phenobarbital, methaqualone, oxazepam, internal standard (dehydroisoandrosterone), diphenylhydantoin, diazepam, chlordiazepoxide, and flurazepam

Although the analysis for ethanol and other low-boiling alcohols outlined above might be considered strictly a dilution-direct injection method, it is included here as one of the extraction methods found suitable for ethanol determination. Other methods of this type will be included in a subsequent section devoted to ethanol analysis in biological fluids by direct injection of sample into the gas chromatograph.

2. Distillation Methods [85-87]

Fox [85] offers an alternative means of separating ethanol from a measured sample of blood, urine, or tissue homogenate for introduction into the gas chromatograph, based on a distillation of the specimen to effect the isolation. Using a Perkin-Elmer model 154-B chromatograph equipped with a thermistor thermal conductivity detector, two columns were used: a 100-cm x 0.64-cm-o.d. copper U-tube packed with glycerol-tricresyl phosphate-C-22 firebrick support (30-60 mesh) prepared in a 30:20:50 ratio (w/w), and a second 100-cm x 0.64-cm-o.d. copper U-tube packed with glycerol-tricresyl phosphate-C-22 firebrick support (30-60 mesh) prepared in an 18:22:60 ratio (w/w). With the helium carrier gas flow rate set at 20 ml/min, the temperature of the first column is maintained at 106°C and that of the second at 84°C to achieve maximum sensitivity.

For analysis, the sample is prepared by introducing a measured amount of blood, urine, or tissue homogenate (1 to 5 g) into a 250-ml distilling flask containing 25 ml of distilled water, 3 ml of 50% sodium tungstate solution, and 5 ml of 1.0 N sulfuric acid. A heated fractionating column. 55 cm by 1 cm in diameter, filled with glass helices, is attached to the flask after 0.25 ml of redistilled n-butanol has been added to the flask's contents. After heating electrically, 20 to 25 ml of distillate is collected in an ice-cooled receiver over a period of 1 hr. A 50- to 100-μl aliquot of the distillate is removed and injected into the chromotograph. Using a reference calibration curve (peak height versus ethanol concentration) prepared by chromatographing blood samples of known alcohol content, quantitative results can be achieved. Although column 1 gave symmetrical ethanol peaks, it did not separate ethanol from either methanol or formaldehyde; column 2, on the other hand, easily resolved these binary mixtures and eliminated the need for performing separate tests for methanol. With the system proposed by Fox, an accuracy of 5 mg% was possible for blood samples containing 0.1 to 3.0 mg of ethanol/ml [89].

Gadsden and McCord [86,87] directed their attention to the development of a preparative method for isolating volatile materials (Fluothane, Fluoromar, alcohols) from biological media. Their approach parallels that of Fox in many respects, consisting of a distillation technique which involves the precipitation of proteins, addition of a water-immiscible solvent, subsequent distillation, followed by dehydration of the distillate [87]. It differs from Fox's method, in which the distillate is chromatographed without regard to its water content. An immiscible solvent is selected on the basis of several criteria: (1) relatively low boiling point, (2) its ability to "carry over" quantitatively the material being analyzed, and (3) separation from the compound(s) to be analyzed must be easily accomplished chromatographically. For the identification and determination of alcohols in blood, the immiscible solvent selected was n-butanol.

A Beckman GC-2 gas chromatograph with a thermal conductivity cell as detector was used in addition to a 25-ft x 1/4-in. copper column packed with a coated solid support (30-60 mesh Chromosorb) similar to that described by Cadman and Johns [59]. Into the proximal end of the tube, anhydrous $CaSO_4$ (30-60 mesh) is packed to a length of 1.5 in. Chromatographic analyses are carried out at 130°C, filament current of 250 mA, and a helium carrier gas inlet pressure of 40 psi.

The procedure for isolating the alcohols from blood consists of the following steps:

Into a round-bottomed distilling flask (Fig. 2.10) containing 4 ml each of sodium tungstate-sodium sulfate and sulfuric acid solutions (these volumes introduced by pipette) and several glass beads or boiling chips, 0.25 ml of n-butyl alcohol is added. A 1-ml aliquot of blood (or other fluid) is pipetted directly into the flask, which is stoppered, swirled vigorously to ensure homogeneous mixing, and then attached to the distilling apparatus. After remaining undisturbed for nearly 5 min or until the precipitated proteins become brownish in color, the flask's contents are gently heated with a flame from a microburner. Into a micro test tube of 1-ml capacity, approximately 0.5 ml of distillate is collected over a period of 2 to 3 min; the upper layer or phase of the distillate consists of n-butyl alcohol and the volatile components under investigation (namely, ethanol). The upper layer is removed and transferred to a small test tube (1-ml capacity) with

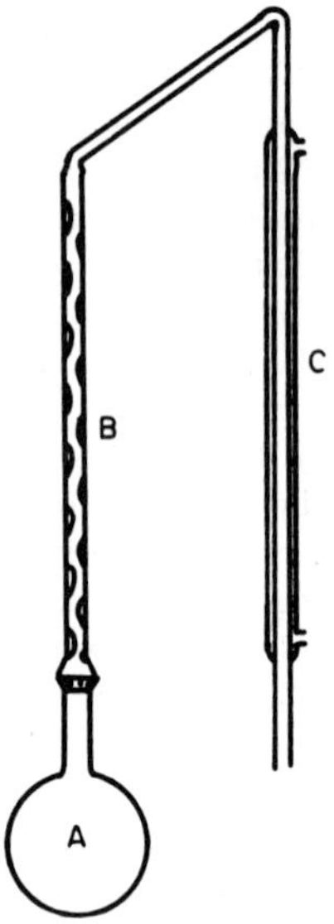

Figure 2.10. Distilling apparatus: Part A-round-bottomed distilling flask, ST 14/20, 100 cm^3 capacity. Part B-Vigreaux column, length 30 cm, i.d. 1 cm. Part C-condenser, length 22 cm, i.d. 0.5 cm. Ball and socket joint between parts B and C gives more mobility to apparatus. From Gadsden and McCord [87], courtesy of Journal of Gas Chromatography.

a micropipette. To this is added 0.1 g of anhydrous sodium sulfate to remove the last traces of water present and, after a standing period of several minutes, an aliquot of the sample is removed and injected into the chromatograph. Based on peak area measurements, reproducibility by the described technique is within ±5%; this is indicated by the values of ethyl alcohol concentration shown in Table 2.7, which were found upon repetitive analyses of samples of known composition. In addition to the identification and analysis of ethanol, the method has been applied to clinical and toxicological problems. For example, it has been possible to identify and determine quantitatively in one sample of 1 ml of blood methyl, ethyl, isopropyl, propyl, and butyl alcohols (as found in the blood and urine of drinkers of illegal whiskey), acetone, paraldehyde, acetaldehyde, and ethyl ether.

3. Direct Injection Methods [88,90-120]

There are quite a number of direct injection methods for the gas chromatographic determination of ethanol and other volatile components in biological media. These methods fall essentially into one of three categories:

TABLE 2.7

Values of Ethyl Alcohol Concentration Found upon Repetitive Analyses of Samples of Known Composition[a]

Ethanol added (mg%)	Ethanol found (mg%)
36	34
36	38
36	36
36	39
36	36
72	86
72	70
72	73
72	69
72	62
144	156
144	138
144	142
144	146
144	145
180	184
180	183
180	191
180	169
180	177
216	213
216	211
216	221
216	229
216	231
288	277
288	271
288	285
288	290
288	287

[a]From McCord and Gadsden [86], courtesy of Journal of Chromatographic Science.

(1) direct injection, (2) precipitation-injection, or (3) dilution-injection. However, in this section, all will be considered as being the same, that is, direct-injection gas chromatographic procedures.

In 1959, Maricq and Molle [90,91] employed gas chromatography for simultaneous determination of diethyl ether and ethanol in blood. This technique was subsequently extended to these volatile components in brain tissue and stomach contents [91]. The analyses were performed with a Perkin-Elmer model 116 chromatograph, using a 400-cm x 0.44-cm copper U-tube column packed with 20% (w/w) Carbowax 1500 on Celite which was operated at a column temperature of 106°C. Using the thermistor-equipped instrument with a hydrogen carrier gas flow rate of 25-ml/min, a 45-min analysis time was required to flush out the water which was eluted after the components of interest.

For quantitative ethanol studies, 1 ml of preserved blood is placed in a microcentrifuge tube, and phosphotungstic acid solution (0.1 ml) containing 150 g of solid/100 ml of water is added. After the mixture is homogenized and centrifuged for nearly 15 min at 2500 to 3000 rpm to precipitate the blood proteins, an aliquot (10 μl) of the supernatant is injected into the chromatograph. With a blood sample containing 0.15% ethanol as a standard, results are calculated by comparative peak area measurements [90].

For simultaneous diethyl ether-ethanol determinations, the blood sample is prepared for analysis by centrifuging 5 ml of blood to which 1 ml of saturated aqueous phosphotungstic acid had been added; from this 10 μl of the supernatant is withdrawn for analysis. A standard is prepared by similarly treating horse blood to which prescribed amounts of ethanol and ethyl ether are added. To determine these components in brain tissue, a prior distillation is required. For example, the chromatographic brain tissue sample was obtained by adding saturated picric acid to 150 g of tissue, distilling the mixture, collecting the distillate in a receiver maintained at 0°C, concentrating the distillate to 10 ml, and finally injecting 10 μl of the distillate into the chromatographic apparatus for analysis. In their procedure, the ethanol-diethyl ether levels in the stomach were established using the same brain tissue distillation-concentration-injection method.

For ethanol in body fluids and tissue, the injection block of an Aerograph model A-110-C chromatograph was modified to permit the insertion of a rolled SS screen into the wide bore of the interchangeable injector [92]. The screen, which functions as a trap for any solid blood particles without restricting the flow of helium carrier gas, is removed after approximately 30 analyses and cleaned with a sodium hypochlorite solution. Having drawn a 50-μl sample of blood into a dilution pipette, 100 μl of a standard solution of aqueous n-propanol is taken up and mixed thoroughly with the blood in the mixing chamber of the pipette. After complete mixing, 10 μl of this diluted blood sample are then injected into the chromatograph equipped with a 300-cm x 0.32-cm-o.d. SS column filled with 30% (w/w) Carbowax 400 on 30-42 mesh C-22 firebrick. With the injector, column, and thermal conductivity detector temperatures held at 260, 108, and 108°C, respectively, and a helium carrier gas flow of 30 ml/min, the elution order is acetone, ethanol, n-propanol, and water, the water having a retention time of nearly 6 min. Using peak height measurements with n-propanol as internal standard, the accuracy was reported to be 1% at an ethanol concentration of 150 mg%.

Chundela and Janak [93,94] simplified the determination of blood ethanol by the use of hydrophilic stationary phases such as Carbowax or diglycerol on Sterchamol as well as an internal standard (n-butanol or methylethylketone). Using a chromatograph equipped with a thermal conductivity cell, they introduced 0.1 ml of blood directly into the instrument (this necessitated a column change after about 15 analyses). Because of the large sample used, methanol and ethanol were not resolved, ethanol values being nearly 8 to 15% lower than those obtained by the Widmark dichromate method. With flame ionization detection, resolution was achieved by direct injection of 0.2 ml of blood. At an ethanol concentration range of 0.8 to 1.9%, the sensitivity and relative error were reported to be 0.1 and ±5%, respectively.

Wesselman [88] developed a routine GC method for ethanol in pharmaceutical preparations and compared his data with the then-current USP IV distillation procedure. A study of available liquid stationary phases (tricresyl phosphate, polyethylene glycol, diglycerol, dinonylphthalate, Ucon 75-H-9000) resulted in the selection of PEG 400 as the best from the point of view of resolution of ethanol, water, and acetone (used as internal standard) and shortness of retention times. A comparison of three sample injectors led to the adoption of the Tenney-Harris system [121] employing 1- to 2-μl Hamilton microsyringes. For routine determinations, standard solutions containing acetone and a suitable amount of ethanol are prepared for each sample (tincture, fluid extract, elixir, or other product) to be assayed. The sample is diluted with the same amount of acetone as is contained in the standard. These solutions (standard and sample) are then chromatographed with a Podbielniak Chromacon using a 6-ft x 1/4-in. copper column packed with 30% (w/w) PEG 400 on 30-60 mesh Chromosorb and operated at 100°C. The other chromatographic conditions selected by Wesselman were injector, 300°C; helium carrier gas flow rate, 75 ml/min; sample size, 1 to 2 μl; TC cell current, 60 mA. Resorting to peak height measurements, comparative data for ethanol in various pharmaceutical preparations by GC and USP distillation are shown in Table 2.8; the GC assay was performed in less than half the time required via distillation.

On the other hand, Drawert et al. [95] resorted to "reaction gas chromatography" in the analysis of alcohols where reaction gas chromatography is defined as a technique for performing reactions in a reactor attached to

TABLE 2.8

Comparison of Ethanol Determinations by Distillation and Gas Chromatography[a]

Preparation	Theory (%)	Distillation (%)	Gas chromatography (%)			
			Analyst 1	Analyst 2	Average	Standard deviation
Elixir of Betalin complex	17.00	17.27	16.90	17.33		
		16.83	17.16	16.66		
		16.67	17.77	17.78	17.27	0.97
F.E. of Cascara sagrada aromatic	18.00	18.00	18.18	17.77		
			17.90	17.85		
			17.50	17.50	17.78	0.52
Elixir of Amytal	34.00	32.88	33.55	33.33		
			34.22	33.60		
			34.10	34.20	33.83	0.91
Fluid of rose, soluble	48.00	49.08	48.88	47.77		
			49.55	48.57		
			48.00	48.71	48.58	1.25
Tincture of belladonna U.S.P.	67.00	67.14	66.66	68.57		
			68.43	66.11		
			68.40	67.14	67.55	2.45
Spirit of camphor N.F.	83.00	86.00	85.33	82.11		
			82.44	81.88		
			83.15	84.64	83.26	3.70

[a]From Wesselman [88], courtesy of Journal of American Pharmaceutical Association.

a GC unit and transferring the reaction products in a closed system to an analytical column. In their initial studies, aqueous solutions of alcohols were injected into a reactor containing a 2:1 (w/w) mixture of Sterchamol: P_2O_5 (previously conditioned for 1 hr at 300°C), with the resulting olefins passed through a second reactor where water is converted to hydrogen gas by reaction with a 1:1 CaH_2:Sterchamol mixture. The olefins dried in this manner are then carried to the analytical column (3.2- or 6.2-m x 6-mm i.d. glass) filled with 26% dinonylphthalate coated on Sterchamol (silica gel) in a stream of hydrogen carrier gas. Located outside the thermostatically controlled enclosure housing the column and TC detector, the reactor in which olefin production occurs is heated separately to 200 to 300°C, while the ranges of the column temperature and carrier gas are maintained between 70 and 120°C and 35 and 50 ml/min, respectively. Using these conditions and a sample size of 0.005 to 0.500 ml, retention data relative to benzene at 100°C are reported for olefins derived from ethanol, n-propanol, four isomeric butanols, four isomeric pentanols, n-hexanol (primary), and n-heptanol (primary).

Applied to the determination of ethanol in blood, ethyl alcohol in the blood sample is first transformed into ethylene in the hot silica gel reactor tube. The ethylene is then separated at 50°C from the other gases in a 3-m x 6-mm-i.d. glass column filled with 65-mesh silica gel-phosphorus pentoxide (50:50 by weight) through which a hydrogen carrier gas flow rate of 16 ml/min is maintained. Showing results that agree rather favorably with those obtained by enzymatic analysis, Drawert and Kupfer [96] also note that the "reaction" analytical method is ten times more sensitive than the technique previously described by Drawert and Kupfer [122], which involves the conversion of alcohols to volatile alkyl nitrites by esterification with nitrous acid.

The method of Machata [97] for blood alcohol determination is based on direct injection of diluted blood or serum specimens into a Perkin-Elmer apparatus equipped with a flame ionization detector. The 2-m Perkin-Elmer "K" column (packed with polyethylene glycol on Kieselguhr) was operated at either 80 or 100°C, depending on whether or not higher alcohols were present in the sample. With a nitrogen carrier gas flow rate of 62 ml/min and an injection port temperature of 140 to 160°C, a 0.5-ml sample of either serum or blood was mixed with exactly 0.20 ml of aqueous 0.25% (w/w) acetone solution (3.165 ml of acetone/liter of water), the acetone serving as an internal standard. Of this mixture, an aliquot was injected directly into a modified injector block. Quantitative determinations were carried out simply by measurement of the peak areas or by automatic integration and comparison with a standard curve prepared by plotting the ratio of the acetone-ethanol peak areas versus the blood alcohol concentration.

In 1962, Parker et al. [98] described a GC method which is applicable to the medico-legal determination of the ethyl alcohol content of blood and aqueous solutions and offers several advantages over other procedures:

namely, rapid analysis, improved accuracy, simplicity, and specificity. Their GC instrument, a HyFi equipped with a flame detector, employed a 10-ft x 1/8-in.-o.d. SS column packed with 60-80 mesh Chromosorb W, acid-washed, coated with castor wax (40% by weight). The operating conditions were as follows: injector temperature, 170°C; column temperature, 120°C; flow rate of nitrogen carrier gas, 13.6 ml/min; hydrogen flow rate, 22 ml/min; sample size, 1 μl. To prepare and quantitate ethyl alcohol in a sample required only 7 min, and the use of ethyl acetate as an internal standard obviated the necessity of precise measurement of blood samples. They reported that standard blood and water samples containing alcohol at concentrations ranging from 0.023 to 0.180% could be analyzed with a precision of 2% and an accuracy of about 4%. Figure 2.11 is a typical chromatogram resulting from 1 μl of prepared blood sample containing 1.0 ml each of ethyl alcohol (0.065%) and ethyl acetate (0.169%). Table 2.9 lists retention data for 56 compounds relative to the retention time (4.2 min) of ethanol; these compounds were studied to investigate the specificity for ethyl alcohol and to estimate the general utility of the method for presumptive identification of other volatiles. The water response (A in Fig. 2.11), in part due to the emerging water which cools the flame detector, was typical of the results obtained from water, urine, or blood injections.

For establishing the ethanol concentration in urine, Bonnichsen and Linturi [99] injected 2 μl of unaltered urine specimens directly into a Perkin-Elmer chromatograph with a flame ionization detector; separation was achieved with a column prepared using 10 g of 10% Carbowax 1500 on Teflon followed by 5 g of 20% Ucon LB-550-X on Chromosorb with temperature held at 100°C.

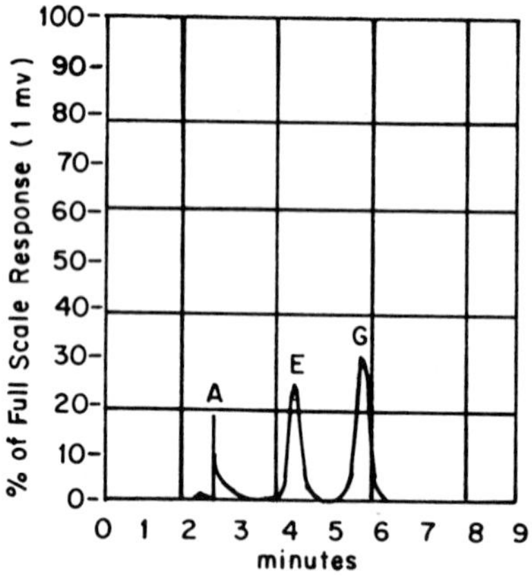

Figure 2.11. Chromatogram of 1 μl of prepared blood sample: 1.00 ml each of ethyl alcohol (0.065%) and ethyl acetate (0.169%). From Parker et al. [98], courtesy of Analytical Chemistry.

TABLE 2.9

Relative Retention Time of Compounds Studied[a]

Compound	Relative retention time	Compound	Relative retention time	Compound	Relative retention time
Acetaldehyde	0.49	Ethyl acetate	1.30	Ethylene dichloride	2.11
n-Pentane	0.54	n-Heptane	1.41	Isobutyl alcohol	2.61
Ethyl ether	0.59	Cyclohexane	1.45	n-Butyronitrile	2.70
Water	0.59	2-Butanone	1.47	Dioxane	3.32
Methanol	0.74	Methyl iodide	1.54	Paraldehyde	3.39
n-Hexane	0.81	Trichloroethane	1.58	n-Butyl alcohol	3.46
Propionaldehyde	0.82	Propionitrile	1.59	2-Ethylbutyraldehyde	3.59
Acrolein	0.83	Ethyl iodide	1.60	Toluene	3.62
Acetone	0.87	Diethylamine	1.62	n-Butyl acetate	4.30
Isopropyl ether	0.89	Carbon tetrachloride	1.67	Amyl alcohol	4.53
Methyl acetate	0.90	Chloroform	1.78	n-Butyl ether	4.74
Methylene chloride	0.98	n-Propyl alcohol	1.80	Ethylene chlorohydrin	5.22
Ethyl alcohol	1.00[b]	Allyl alcohol	1.88	Pyridine	6.02
Acetonitrile	1.05	Allyl ether	1.96	Chlorobenzene	6.94
Acrylonitrile	1.13	Benzene	1.98	Isoamyl alcohol	7.38
Isopropyl alcohol	1.15	Acetylene dichloride	2.04	Formic acid	c
n-Butylamine	1.15	sec-Butyl alcohol	2.04	Acetic acid	c
tert-Butyl alcohol	1.22	Ethylene trichloride	2.09	Lactic acid	c
				Formaldehyde	c
				Carbon disulfide	c

[a]From Parker et al. [98], courtesy of Analytical Chemistry.
[b]Ethyl alcohol retention time 4.2 min.
[c]0.2 μl of compound injected and no response observed.

On the other hand, in 1963, Coldwell and Grant [100] performed duplicate analyses on 36 samples of blood and 35 urine specimens, using a modification of the GC method of Cadman and Johns [59] and the desiccation procedure proposed by Smith [123]; the alcohol levels ranged from 1 to 81 mg% (averaging about 40 mg%). The mean difference between duplicate analyses by GC was 2.5 for blood and 1.1 mg% for urine; the standard error for blood and urine was 2.05 and 1.32 mg%, respectively. With the Smith procedure, the mean difference and standard error for blood were 3.1 and 2.12 mg%, respectively, whereas for urine, the respective values obtained for the data were 3.9 and 1.48 mg%. As noted by the authors, for a given sample of blood, the maximum deviation between the GC and Smith methods ranged from +12.4 to -12.4 mg%. The following substances were separated and estimated quantitatively: aldehydes (acetaldehyde and paraldehyde), acetone, ethanol, ethyl ether, methanol, isopropanol, and n-propanol. Furthermore, they showed that tissue homogenized with 5 N KOH could be analyzed directly for alcohol, thereby eliminating the need for removing the alcohol by distillation.

Using a flame-equipped chromatograph (F&M model 400), Mather and Assimos [104] selected a column packed with 20% Hallcomid on acid-washed, 60-80 mesh Diaptoport W which was operated at 70°C. For quantitative results, the injector and detector temperatures were maintained at 180 and 110°C, respectively. Helium was used as carrier gas, and its flow rate was held at 60 to 80 ml/min. Sample preparation was rather simple and straightforward: 2 ml of a blood, plasma, or serum specimen was pipetted into a small, capped plastic tube and 0.2 ml of 1% dioxane (internal standard) was then added, followed by thorough mixing. The ethanol concentration was estimated from the measured peak heights obtained for the aliquot injected directly into the chromatograph by comparison to a standard curve prepared by plotting peak height ratios against blood alcohol concentrations.

In 1966, the procedure proposed by Curry et al. [80,106] permitted the ethanol analysis to be performed with sufficient accuracy, precision, and speed that it became accepted for routine use in the United Kingdom. Applicable to small blood samples, the blood sample (20 μl) is diluted with 100 μl of aqueous n-propanol solution with a Griffin and George type 221 hemoglobin-type Diluspence; the propanol serves as internal standard. Approximately 1 μl of the diluted blood sample is injected by means of a Hamilton syringe into the chromatograph. The chromatograph employed was a Pye model 104 with a 5-ft, 10% PEG 400 on 100-200 mesh Celite column and a flame ionization detector, both operated at a temperature of 85°C in addition to its injection port. With the argon (carrier gas), hydrogen, and air flow rates set at 150, 100, and 600 ml/min, respectively, ethanol had a retention time of 2 min whereas propanol and the "water" peak were eluted in 3.25 and 4.50 min, respectively. Calibration, and hence measurement of concentrations, was achieved by plotting the ratio of ethanol-to-propanol integrator signals against ethanol concentration in the blood. The graph obtained was independent of the volume of blood injected and, when 20 injections of the same solution were made (250 mg/100 ml), the standard deviation for the ethanol-to-propanol ratio, having a mean value of 0.7924, was 0.00580. They noted that the column would easily resolve ether, acetaldehyde, acetone, methyl ethyl ketone, and methanol from ethanol; however, isopropanol, having only a slightly shorter retention time than the ethanol peak under these circumstances, became distorted. This difficulty can be overcome if one uses a column containing Porapak Q at 185°C. In this case, both alcohols are completely resolved.

To circumvent some of the difficulties encountered because nonvolatile sample components may contaminate the column or interfere with detection, Freund [107] designed an exchangeable injection port cartridge for the column entrance which permitted direct on-column injection of biological samples (Fig. 2.12). The apparatus used was an F&M model 402 gas chromatograph equipped with a hydrogen flame detector and a 5-ft-long, 0.125-in.-o.d. SS column packed with 150-200 mesh Porapak Q. The operating conditions employed for blood, urine, or aqueous standards were as follows: injection port temperature, 240°C; column temperature, 115°C; detector temperature, 180°C; flow rates: nitrogen (carrier gas), 15 ml/min; hydrogen, 25 ml/min; air, 350 ml/min. Using a 10-μl syringe, 2 μl of blood, urine, or aqueous standard were injected into the interchangeable injection port following a prescribed aspiration procedure. The recovery of acetone in aqueous solution was 99.6 ± 2.6%.

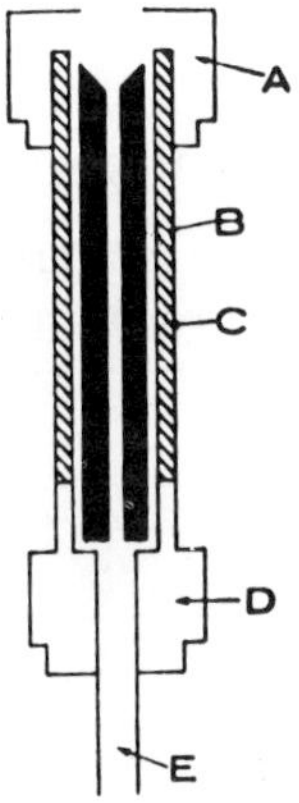

Figure 2.12. Injection port. A, Swagelok tube fitting, 0.25 in. o.d.; B, cartridge; C, stainless-steel tubing, 0.25 in.; D, Swagelok reducer, 0.25 to 0.125 in.; E, column, 0.125 in. o.d. From Freund [107], courtesy of Analytical Chemistry.

For the determination of blood ethanol by gas chromatography, Colehour [108] used water as an internal standard. For all determinations, an F&M model 402 GC instrument with a flame detector was used, its U-shaped glass column, 66 cm long and 4 mm i.d., filled with 50-80 mesh Porapak Q. After conditioning the column with repeated injections of 40% methyl amine, two peaks appeared in the chromatogram when 0.1 μl of 250 mg% of ethanol standard in water was injected. The second peak appearing early in the chromatogram was attributed to water. Its area remained constant in relation to the ethanol area even after several 0.1-μl injections of the standard; therefore, it was used as the internal reference. Colehour used the following procedure for preparing blood specimens for analysis:

> Capillary blood samples were collected in nonheparinized microhematrocrit tubes, capped, and allowed to clot. The samples were centrifuged; 0.9 microliter of air and then 0.1 microliter of serum were drawn into the syringe by putting the needle directly in the capillary. The sample was injected into the chromatograph as quickly as possible. Two or more ethanol-water samples of similar percentage composition were also injected alternately with the serum. Swept areas were measured as described above; alcohol-aqueous area ratios were plotted for the standards, the unknowns were similarly measured, and the percentage of ethanol read off the curve. If increased accuracy is required, duplicate determination may be made.

Table 2.10 lists the recovery of absolute amounts of ethanol in the blood at various concentrations by the proposed method, whereas Figure 2.13 is a composite chromatogram of several sera prepared from bloods containing different levels of ethanol. The theoretical basis for this analysis is the fact that the column is partially loaded with strongly basic methyl

TABLE 2.10

Recovery of Absolute Amounts of Ethanol from Blood by Proposed Method[a]

Ethanol added (mg%)	Ethanol found (mg%)[b]
25.0	27.8 ± 1.5
75.0	78.3 ± 3.3
150.0	147.2 ± 11.6
250.0	249.2 ± 13.7

[a]From Colehour [108], courtesy of Analytical Chemistry.
[b]Each value is a mean of six single determinations.

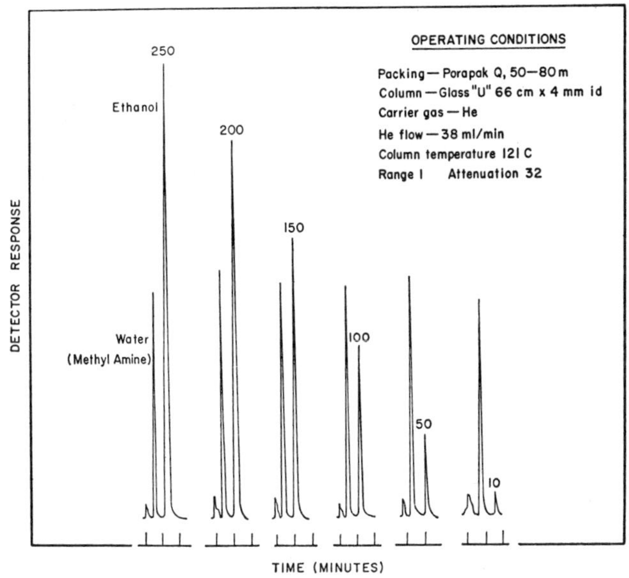

Figure 2.13. Chromatogram of blood samples containing various amounts of ethanol using water as the internal standard. From Colehour [108], courtesy of Analytical Chemistry.

amine, a portion of which is eluted quantitatively by the water in the standard or sample depending on the load. With time, water response gradually diminishes as the amine is depleted. Therefore, it must be replenished but, as pointed out by Colehour, this diminution is kept below a significant level by the use of small samples (0.1 μl).

Rather than construct a special, removable injection port, Cooper [109] overcame the tendency of blood coagulating and blocking the syringe on entering the injection port by precipitating (in a way similar to the work of Maricq and Molle) the proteins with a mixture of sodium tungstate and sulfuric acid. A Pye 104 gas chromatograph was used with a 5-ft Porapak Q column and flame ionization detector. The operating conditions were as follows: column temperature, 170°C; detector temperature, 200°C; nitrogen carrier gas flow rate, 50 ml/min. For analysis, the blood sample was prepared in the following manner: 0.2 ml of blood was diluted with 3.0 ml of aqueous n-propanol solution (10 mg/100 ml of water, which is suitable for ethanol concentrations in the 50-to-250 mg/100 ml range). To this was added 0.15 ml of aqueous 10% sodium tungstate and 0.15 ml of 0.67 N sulfuric acid. After centrifuging, 5 μl of the supernatant was injected onto the column. Submitted to GC analysis, known ethanol concentrations added to blood gave recoveries of 90 to 104% and, for 20 separate dilutions of blood containing 150 mg/100 ml of added ethanol, the standard deviation and coefficient of variation for this study were 2.9 mg/100 ml and 2.0%, respectively.

Both Finkle [111] and Jain [110] also used gas chromatographs equipped with flame ionization detectors. In the method of Finkle, a glass column was packed with 5% Hallcomid M-18 and 0.5% Carbowax 600 on 40-60 mesh Teflon 6. The column temperature was set at 50°C with the detector and injection port maintained at 75 and 150°C, respectively. Following a 1-μl water injection as a blank control, a 0.5-μl sample of ethanol reference solution (200 μl/ml) was injected. This was followed by a 0.5-μl injection of plasma serum or urine. A second standard of mixed volatiles was used if a peak other than ethanol occurred; this second standard solution contained acetaldehyde, acetone, ethanol, isopropanol, methanol, paraldehyde, toluene, and trichloroethanol. By comparing the area under each peak with that of one in the mixed standard which had a similar retention time, the concentration of each eluted component could be estimated.

In his GC method, Jain used a Varian Hy-Fi model 600-D gas chromatograph. Operated at either 100 or 130°C, the column was filled with 30% Carbowax 20M on acid-washed, 60-80 mesh Chromosorb W. At a nitrogen carrier gas flow rate of 35 ml/min, 0.5 μl of known and unknown mixtures were injected into the chromatograph, whose inlet was maintained at 160°C. Sample preparation consisted of setting up two different 2-dram vials (one marked "known" and the other "unknown") into which was pipetted individually 0.5 ml of isobutanol internal standard solution (5 ml of iso-BuOH standard stock solution, 10 mg/ml, diluted to 100 ml with distilled water). Whereas 0.5 ml of the unknown blood was transferred into the "unknown" vial followed by three washings with the internal standard solution to ensure complete transfer of the blood specimen, 0.5 ml of the working ethanol solution (50 mg/100 ml) was added to the vial marked "known." These mixtures were injected into the gas chromatograph. The quantity of ethanol present was determined using the following expression:

$$\text{Ethanol concentration (mg/100 ml)} = C_X = C_S\left(\frac{h_X}{h_S}\right)\left(\frac{H_S}{H_X}\right) \qquad (2.19)$$

where

C_X = ethanol concentration in the blood sample, mg/100 ml
C_S = concentration of the standard alcohol solution, mg/100 ml
h_X = peak height of ethanol from "unknown" blood
h_S = peak height of isobutanol internal standard from "unknown" blood
H_S = peak height of ethanol from "known" ethanol solution
H_X = peak height of isobutanol internal standard from "known" ethanol solution

In 1972, Solon et al. [112] introduced an automated GC method for the analysis of ethanol in either urine or blood. The automated injection system consisted on a Hewlett-Packard blood alcohol analyzer, which in turn comprised a gas chromatographic unit, an automatic liquid sampler, an electronic integrator, a data processor, and special software. The system used the following accessories and operational conditions:

Column: 6-ft x 1/8-in.-i.d. SS, packed with 50% Porapak Q and 50% Porapak R preceded by 10-cm x 3-mm-i.d. glass pre-column filled with glass wool.

Temperatures: Injection port, column, and flame detector: 150, 150, and 300°C, respectively.

Flows: Nitrogen (carrier gas), air, and hydrogen: 50, 350, and 28 ml/min, respectively.

Control module for sampling: Analysis cycle, 8 min; stop integration, 7 min.

Sample preparation: To 1.0 ml of n-propanol internal standard solution (2 ml of n-propanol/liter of water) is added 0.5 ml of blood sample, followed by thorough mixing. The calibration standard is prepared by diluting to 1 liter with distilled water a binary mixture containing 2.0 ml of n-propanol and 20 ml of ethanol.

Instrument operation: Controlled by a subroutine in the data processor software which automatically advances the sampler turntable to each of its 36 positions, positions 1 and 2 of the sampler are filled with vials containing heparin rinse solution. In numerical sequence, the calibrator solution occupies position 3, with the remaining 33 positions filled with urine or blood specimens to be analyzed for their ethanol content. When the sampler is advanced to position 1 by the processor, the syringe is rinsed. Following this operation, it is moved to position 3, where the standard solution is sampled and injected into the chromatograph. With the retention times and response factors relative to n-propanol (internal standard) stored for future or later reference, the sampler then returns to position 1 to rinse the syringe again before it is advanced to position 4 (the location of the first diluted specimen). Here it withdraws a fixed volume of specimen (dictated by the size of the syringe barrel) for injection. Following injection, the data is analyzed and then the cycle described above is repeated until all of the samples have been analyzed sequentially. However, the vial in position 2 is used during the second half of the run for rinsing the syringe.

With this computer-controlled, automated system, Solon et al. reported an overall relative standard deviation equal to ±1.00% over a concentration range between 0.015 and 0.260% for which there existed an excellent linear relationship between GC response and ethanol concentration.

Baird [116] modified the method advocated by Solon et al. by interfacing the Hewlett-Packard 5700 GC and automatic sampling system with an Autolab System IV computing integrator. In addition to the introduction of new hardware features, the sample-internal dilution ratio was increased to 1:8 and analytical results were computed on a weight/volume basis. The accuracy of the method was established by analyzing 25 blood samples which had previously been analyzed by a distillation-oxidimetric procedure and comparing the results. The average standard deviation of results between the two methods was ±0.003% blood alcohol. On the other hand, the precision (based on 400 different urine and blood specimens analyzed and prepared in duplicate) yielded an average deviation for the duplicate values obtained 90 min apart of ±0.003%; the average blood alcohol concentration in the specimens investigated in this study was 0.19%.

Sine and co-workers [114] developed a rapid and uncomplicated method of analysis for a variety of drugs subject to abuse and accidental overdose as listed in Table 2.11 in which such aliphatic alcohols as methanol, ethanol, and isopropanol are included. Preparation of samples involves a simple serum extraction and aqueous dilution technique followed by a GC analysis using Varian Aerograph GC instruments (model 2100 or 1400) equipped with hydrogen flame ionization detectors and glass columns.

The aliphatic alcohols and acetone were determined by an "on-column" technique using a 10-ft x 2-mm-i.d. glass column packed with 7% Hallcomid M-18 coated on Porapak Q and isothermally maintained at 110°C with the injector and detector at 150°C. Optimum carrier gas (nitrogen) flow was determined for this column as described by McNair and Bonelli [124], with the hydrogen and air reaching the detector regulated at 50 and 200 ml/min, respectively.

The recommended procedure for alcohol-acetone analysis includes the following steps:

1. Label two 10-ml glass test tubes, one as unknown (sample) and the other as standard.
2. To the "standard" test tube, accurately pipette 2 ml of standard solution (50 mg each of acetone, ethanol, methanol, and isopropanol/ 100 ml of distilled water) and 1 ml of distilled water.
3. To the test tube labeled unknown or sample, pipette 2 ml of distilled water and 1 ml of the patient's serum.
4. After both test tubes have been gently swirled to ensure complete sample-water mixing, a 10-μl syringe (Hamilton equipped with "Repro-Jector" automatic sample injectors) is flushed several times with sample and 2 μl is then injected onto the Hallcomid column.

TABLE 2.11

The Significance of Drug Blood Levels[a]

Generic name	Trade name	Therapeutic[b] (mg/100 ml)	Toxic[b] (mg/100 ml)
Hypnotics			
Amobarbital	Amytal	0.03-0.5	3-5
		0.1 -0.5	1-3
Barbital	Vernal	0.5 -1.5	7-10
		1	6-8
Butabarbital	Butisol	0.03-0.5	3-5
		0.1 -0.5	1-3
Chloral hydrate	Noctec	1	10
Chlordiazepoxide	Librium	0.25-0.8	2-5
		0.1 -0.2	0.6
Diazepam	Valium	0.1 -0.6	1.5-3
		0.05-0.25	0.5-2
Ethchlorvynol	Placidyl	0.5	2
Flurazepam	Dalmane	-	-
Glutethimide	Doriden	0.05-0.25	2-3
		0.02	1-8
Methyprylon	Noludar	1	3-6
Pentobarbital	Nembutal	0.01-0.1	1-3
		0.1	0.7
Phenobarbital	Luminal	0.5 -1.5	7-10
		1	4-6
Secobarbital	Seconal	0.01-0.1	1-3
		0.1	0.7
Analgesics			
Meperidine	Demerol	0.01-0.1	1-3
		0.065	0.5
Pentazocine	Talwin	0.015	-
		0.06	0.5
Psychotherapeutic			
Amitriptyline	Elavil	0.01-0.03	0.1-0.5
		-	0.04
Chlorpromazine	Thorazine	0.01-0.05	0.1-0.5
		0.05	0.1-0.2
Imipramine	Tofranil	0.01-0.05	0.1-0.5
Promazine	Sparine	0.01-0.05	0.1-0.5

TABLE 2.11 (continued)

Generic name	Trade name	Therapeutic[b] (mg/100 ml)	Toxic[b] (mg/100 ml)
Anticonvulsants			
Diphenylhydantoin	Dilantin	0.8-1.5	8-10
		0.6-1.7	2-5
Primidone	Mysoline	-	9.3
Aliphatic alcohols			
Ethanol	-	5-20	400-500
		-	150
Isopropanol	-	0	350
Methanol	-	0	75-125
		0	80

[a]From Sine et al. [114], courtesy of Journal of Chromatographic Science.

[b]Data compiled from Ladd and Wallace [125], Winek [126], Sunshine [127], and Clarke [128].

Using this aqueous dilution technique for sample preparation, the reproducibility of the method (given as relative standard deviation) for acetone, ethanol, isopropanol, and methanol was 2.2, 3.2, 1.5, and 5.2%, respectively. A typical GC chromatogram of the standard mixture of alcohols and acetone is shown in Figure 2.14.

Using an on-line computer which continuously acquired and stored information in addition to controlling the scan rate of the LKB mass spectrometer, Bonnichsen et al. [113] determined ethanol using an integrated GC-MS-COMP LKB 9000 system; the quantitative values were based on the ion current generated by its molecular ion (M = 46). The integrated GC-MS unit was equipped with a 1.5-m x 2-mm column packed with 20% Polywax 4000. Sampled at a frequency of 8 to 12 kc, the electron multiplier's signal is fed to the analog input of an IBM 1800 computer. By monitoring the retention time for ethanol and the intensity ratio of m/e 31 ($-CH_2OH$) to m/e 45 (M-1) and m/e 46 (M), proper instrument operation could be assessed at all times. At a column temperature of 90°C, ethanol is eluted in approximately 70 sec following the injection of a 3 μl sample whose alcohol concentration could range from 0.01 to 0.40%. Using three ethanol standards to cover the concentration range normally encountered in unknown samples, the retention time for ethanol is specified. Stored in memory together with the five highest peaks appearing above m/e 28 are the intensities of the m/e peaks 29, 31, 32, 43, 44, 45, 46, and 48 (these

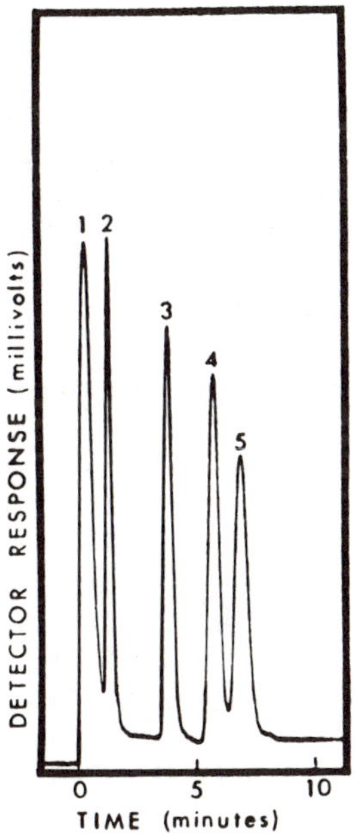

Figure 2.14. A representative GLC chromatogram of the standard mixture of aliphatic alcohols and acetone: (1) solvent front (water), (2) methanol, (3) ethanol, (4) acetone, and (5) isopropanol. (Column: 7% Hallcomid M-18 on Porapak Q.) From Sine et al. [114], courtesy of Journal of Chromatographic Science.

may or may not be included as part of the five most intense peaks noted). In their GC-MS-COMP procedure, Bonnichsen et al. use peaks other than m/e 31, 45, and 46 to establish the presence of other components that might be present. After a maximum intensity is recorded for each component, the computer starts a scan by giving out a pulse and also simultaneously starts to determine the retention time for various components at the onset of sample injection. Ethyl alcohol evaluation commences when a scan stops, and if all conditions set forth in the program are met or fulfilled, a message is printed out identifying this spectrum as ethyl alcohol or a mixture followed by the sample's ethanol concentration as determined by analysis. For this type of instrumental arrangement, an accuracy of better than ±5% is possible for alcohol samples whose concentrations range from 0.02 to 0.30%.

In the absence of more sophisticated instrumentation such as a mass spectrometer, Blume et al. [115] identified the presence of ethanol in blood or other biological specimens by demonstrating the disappearance of the associated peak following treatment of the specimen with alcohol dehydrogenase (this approach was taken for the identification of the unknown as ethanol which is generally based on its retention time).

As noted by the authors, quantitative GC data for ethanol levels in blood or other media were obtained using a procedure similar to that described by Jain [110] insofar as the specimen is simply diluted with an internal standard solution and introduced directly into the chromatograph equipped with a flame ionization detector. Blume et al., however, preferred for their study a 6-ft x 1/8-in.-o.d. SS column packed with 80-100 mesh Porapak S (maintained isothermally at 200°C) and the use of methyl ethyl ketone as internal standard.

For the enzymatic confirmation of ethanol, they prepared a 1 mg/ml solution of yeast alcohol dehydrogenase with deionized water. Ten milligrams of NAD are added to a small test tube containing 1 ml of pyrophosphate buffer solution (sodium pyrophosphate, 0.075 M; semicarbazide, 0.075 M; and glycine, 0.022 M). After thorough mixing of the NAD-pyrophosphate solution, 0.20 ml of alcohol dehydrogenase solution is added. Again, the test tube's contents are mixed. This is followed by the addition of 0.20 ml of the 1:1 dilution of specimen (1.00 ml) in internal standard solution (1.00 ml). After the tube has been stoppered, mixed by inversion, and placed for 30 min in a 37°C water bath, a 0.5-μl aliquot is injected into the chromatograph, whose detector sensitivity is about ten times greater than that normally used to obtain quantitative data. Visually, one can clearly see the almost complete disappearance of the peak associated with ethanol (Fig. 2.15), but the reduction in the size of the MEK peak is unfortunately the consequence of using semicarbazide as a trapping agent, which is necessary to achieve significant oxidation of the ethanol because the equilibrium of the basic reaction does not favor acetaldehyde production.

In 1973, Perkin-Elmer introduced a sampling system (syringeless), the MS-41 capsule sampler, in which each sample to be analyzed is placed in an aluminum capsule by any convenient means; the capsule (3 mm in diameter and 6 mm in length with a maximum volume of nearly 20 μl after sealing) is weighed before and after filling to determine the sample weight. For the analysis of high-boiling components, the solvent may be evaporated before the container is sealed by cold-welding. Placed firmly into a numbered magazine (either for immediate analysis or for storage), the automatic sampling system is so designed that, without any further manual manipulation, a probe can pick up the capsule from the magazine for insertion into the GC injector. After flushing with carrier gas to remove any atmosphere that might be inadvertently introduced with the probe, the capsule is heated to injection temperature, opened by piercing with a hollow needle, and swept free of its volatile components down the hollow needle by the carrier gas under pressure and onto the chromatographic column. The capsule is automatically removed from the injection port after each analysis, taking with it the residual solids that contaminate conventional injection ports and columns.

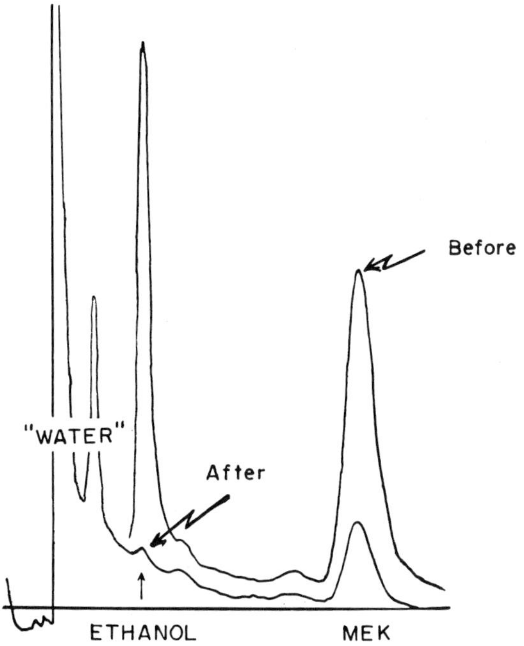

Figure 2.15. Chromatograms of the same reaction mixture before and after alcohol dehydrogenase addition and incubation. Adopted from Blume et al. [115].

Hartigan and co-workers [117] demonstrated its versatility with the analysis of cough drop volatiles (eucalyptol, benzyl alcohol, camphor, menthol, and thymol) by the direct introduction of a piece of cough drop without any sample pretreatment. Chromatographic analsis was completed in less than 5 min using a 50-ft OV-101/graphite support-coated open-tubular (SCOT) column. More appropriately, however, it is possible to introduce blood samples directly into a gas chromatograph for the determination of blood alcohol; the results of such an injection are shown in Figure 2.16 for a sample that required no time-consuming sample pretreatment.

Using GC to monitor the stability of ethanol in stored blood, Brown et al. [118] studied five variables or factors--time, fluoride concentration, alcohol concentration, temperature, and type of container--by means of a factorial experiment. With adaptations of the method described by Curry et al. [80,106] and n-propanol used as internal standard, the gas chromatographs used were a Perkin-Elmer F-11 and a Perkin-Elmer Multifract F-40 headspace analyzer; the instrumental conditions employed are listed below together with the variables in the factorial experiment.

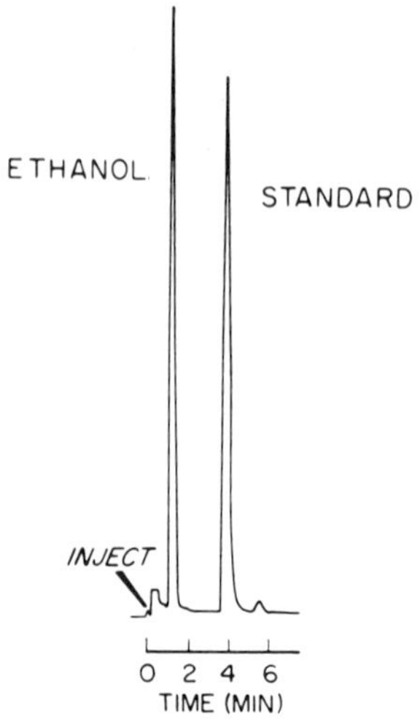

Figure 2.16. Determination of blood alcohol by direct introduction of blood samples. From Hartigan et al. [117], courtesy of International Scientific Communications, Inc.

Instrumental Conditions

Perkin Elmer F11	
Column	4 ft Porapak Q
Oven temperature	170°C
Injection volume	1.5 μl
Carrier gas	Nitrogen
Detector	Flame ionization
Perkin Elmer F40 Autoanalyzer	
Column	5 ft Porapak Q
Oven temperature	140°C
Water bath temperature	62°C
Injection time	5 sec
Carrier gas	Nitrogen
Detector	Flame ionization

Note: Propanol solutions used with the F40 Autoanalyzer also contained 0.5% (w/v) sodium hydrogen sulphite.

Variables of the Factorial Experiment

Time of storage	4 weeks and 8 weeks
Fluoride concentration	Nil and 1% (w/v) NaF
Alcohol level	110 mg% and 220 mg%
Temperature of storage	4 and 17°C
Container	Polypropylene cups and sealed, ring-snap, glass ampoules

As a consequence of their undertaking, Brown et al. summarized their findings as follows:

> The effects of the factors time, sodium fluoride concentration, ethanol concentration, temperature of storage and type of container on the ethanol losses from stored human blood have been investigated by means of a 2^5 factorial experiment. The important factors were found to be temperature, fluoride concentration and time of storage. A detailed study of the important factors enabled three distinct mechanisms of ethanol loss to be identified. These were a highly temperature-dependent oxidation reaction which was independent of the ethanol concentration over a wide range; destruction of ethanol by the action of micro-organisms in the absence of a preservative, which could be inhibited by 0.5% (w/v) sodium fluoride, and diffusion which was found to occur from 5.6% of the polypropylene containers used in Britain for the purposes of the Road Traffic Act 1972.

Smalldon and Brown [119] investigated further the oxidation of ethanol in stored human blood and showed that the oxidizing activity arises from an oxyhemoglobin intermediate and can be inhibited by compounds such as sodium azide, hydrogensulfite, dithionite, and nitrite, which destroy oxyhemoglobin. The reaction is first-order with respect to oxyhemoglobin and zero-order with respect to ethanol concentration. From an Arrhenius plot ($\log_{10}k$ versus $1/T$, where k is the reaction rate and T is the temperature), the activation energy for the production of the intermediate was calculated to be 25 kcal/mole. Proposing an overall mechanism for blood ethanol oxidation, Smalldon and Brown discuss in detail the stabilization of blood ethanol solutions, indicating that sodium azide as a preservative would appear to show considerable promise in this respect. However, they note that further work is clearly necessary to select the inhibitor and its concentration which best fulfills the above criteria and provides stable blood ethanol solutions.

Pereira et al. [120] in 1974 described a specific, sensitive, rapid, and reliable method for the quantitative determination of ethanol in blood and urine by mass fragmentography. The instrumentation and operating conditions employed for their investigation were as follows.

A Finnigan 1015 quadrupole mass spectrometer was coupled by an all-glass membrane separator to a Varian model 1200 gas chromatograph, interfaced to the ACME computer system of the Stanford University Medical School [129]. The GC separations were performed with a 6-ft x 1/8-in.-i.d. coiled glass column packed with Porapak Q, 80-100 mesh, a helium carrier gas flow rate of 50 ml/min, and an isothermal column temperature of 170°C.

To establish that mass fragmentography was indeed feasible, a subject consumed during a 20-min period 40 g of ethanol mixed with 150 ml of orange juice. Blood samples (200 μl in volume), withdrawn periodically from the antecubital vein, were mixed with 90 μl of the deuterated ($^2H_5$) ethanol standard solution (internal standard prepared by diluting 30 μl of $^2H_6$-ethanol with distilled water to 25 ml, the $C_2{}^2H_5O^2H$ being converted into $C_2{}^2H_5OH$ under these conditions) and 1- to 2-μl aliquots injected into the gas chromatograph. In addition to the blood samples, urine specimens were taken from the same subject periodically and analyzed by the method outlined above; for both types of sample, however, data acquisition was commenced 80 sec after sample injection. Calibration curves were prepared by mixing in separate vials increasing amounts of nondeuterated ethanol (10, 25, 50, 100, 150, 250, and 500 μl) standard solution with 90 μl of deuterated ethanol standard solution. The samples were mixed, each being processed by the GC-MS-COMP system, which plotted the ratio of the peak areas of the selected fragment ions of undeuterated (m/e 31 [$CH_2{=}\overset{+}{O}H$] and 45 [$CH_3CH{=}\overset{+}{O}H$]) and deuterated (m/e 33 [$C^2H_2{=}\overset{+}{O}H$] and 49 [$C^3H_3C^2H{=}\overset{+}{O}H$]) ethanol against the amount of ethanol added, that is, the ratio of m/e 45:49 or m/e 31:33 versus ethanol added.

Using this technique, Figure 2.17 represents the ethanol decay profiles in blood and urine after oral ingestion of ethanol (40 g)-orange juice (150 ml) mixture. Realizing that the GC-MS-COMP system described is fairly expensive to use routinely for ethanol analysis, the authors are quick to point out, and rightly so, that the procedure is specific and capable of quantitating as little as 5 ng of ethanol. The percent standard deviation over five values was 3.7% for m/e 31:33 and 1.5% for m/e 45:49.

4. Headspace Methods [130-153]

In 1964, Goldbaum, Domanski, and Schloegel [133] described a GC procedure for the analysis of volatile compounds in biological specimens that depends on the rapid partition that takes place between a volatile compound in solution and the atmosphere (or air) above it. This method is based on the findings of Harger et al. [154], who investigated the partition of alcohol between water, blood, urine, and air and then demonstrated the feasibility of determining the alcohol concentration in the liquid phase from air sample analyses where the concentration of alcohol in the air phase is independent of the volume of the solution, but is primarily influenced by temperature and salt concentration.

The instrument used for the determination of volatiles was a Chromolab gas chromatograph equipped with a beta ionization (radium 226) microdetector and a 6-ft x 3-mm-i.d. borosilicate coiled glass tube packed with a mixed liquid stationary phase [PEG 400, diisodecyl phthalate, and Flavol 8N8, which is 2,2'-(2-ethylhexamido)diethyl di-(2-ethylhexoate)] coated on

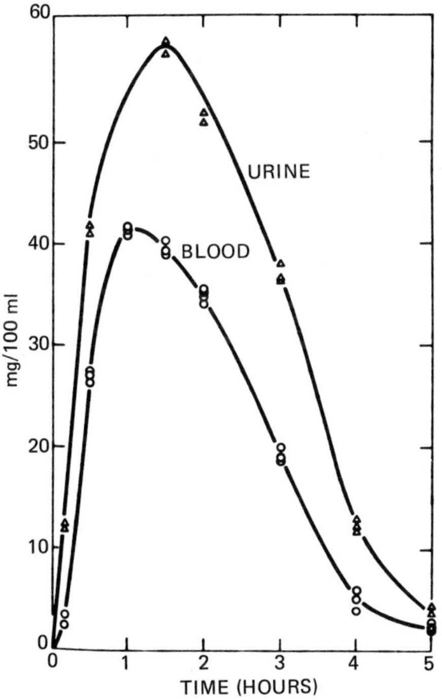

Figure 2.17. Blood and urine ethanol decay profiles for subject after ingesting an ethanol-orange juice mixture. From Pereira et al. [120], courtesy of Clinica Chimica Acta.

42-60 mesh C-22 firebrick, a packing that was first described by Cadman and Johns [59] in 1958. With argon as carrier gas, the GC operating conditions were as follows: column temperature, 95 ± 5°C, injector temperature, 105°C; detector temperature, 150°C; relative gain, 1 x 10^{-8} A.

Using containers stoppered with a puncture-type rubber cap, the unknown specimens (biological fluids or postmortem samples such as liver, brain, or other organs) and standards are placed in a 25°C water bath until temperature equilibrium is reached, that is, until the contents of the containers reach 25°C. Using an airtight syringe, 1 ml of equilibrated air sample is removed for analysis. After disengaging the needle from the syringe, the syringe is then attached to a two-way surgical stopcock which, in turn, is attached to the GC unit with a 19-gauge needle penetrating the half-hole silicone rubber stopper located at the top on the chromatographic column. The two-way stopcock is opened and the sample is injected into the heated inlet of the chromatograph. After injection, the two-way stopcock is closed and the air sample, which appears as a negative peak, marks the point from which retention times are determined. At the above-specified operating conditions, the retention times of compounds of toxicological interest relative to that of ethanol (ret. time = 2.5 min) are listed in Table 2.12; all compounds eluted in less than 30 min after sample injection, in addition to many other compounds having boiling points less than 150°C. In addition to the determination of blood ethanol by this method, Goldbaum et al. were able to identify compounds which yield more volatile components by means of chemical reactions. For example, chloral hydrate can be quantitatively hydrolyzed to chloroform with alkali, and paraldehyde is converted to acetaldehyde in an acid medium.

TABLE 2.12

Relative Retention Time of Volatile Compounds Studied[a]

Compound	Relative retention time	Compound	Relative retention time
Acetaldehyde	0.33	n-Propanol	2.25
Ethyl ether	0.33	Chloroform	2.53
n-Hexane	0.53	Trichloroethylene	2.59
Formaldehyde	0.56	Propyl acetate	2.64
Acetone	0.72	Ethylene dichloride	2.86
Methyl alcohol	0.89	Isobutyl alcohol	3.53
Ethyl alcohol	1.00	Toluene	4.19
n-Heptane	1.05	Tetrachloroethylene	4.41
Isopropyl alcohol	1.19	n-Butyl alcohol	4.86
Methyl ethyl ketone	1.44	Paraldehyde	4.97
Ethyl acetate	1.49	Amyl acetate	8.03
Carbon tetrachloride	1.61	Xylene	8.55
Benzene	2.08		

[a]From Goldbaum et al. [133], courtesy of Journal of Forensic Sciences.

Natelson and Stellate [134,135] devised a chamber for sampling head-space volatiles which is used for the routine assay of ethanol, methanol, acetone, and isopropanol in body fluids. For the gas chromatographic analysis of blood volatiles, 0.1 ml of blood is injected into a 5-ml chamber (supported on a Vortex Jr. mixer in a heating block at 50°C) containing 250 mg of anhydrous copper sulfate spread as a separate layer over 5 g of perforated, 3-mm-diameter glass beads. After the chamber is purged with helium and isolated, the sample is injected. The vibration of the Vortex mixer is sufficient to permit the dehydration of the sample by the anhydrous $CuSO_4$ and the liberation of ethanol and other volatiles which are subsequently swept by a stream of carrier gas onto the gas chromatographic column.

With thermistor detectors, two columns arranged in series were used: a 15-ft x 0.085-in. column packed with Hallcomid M-18 (a mixture of N,N-dimethylamides of myristic, stearic, and oleic acids in the ratio 5:50:5) on 35-mesh Teflon 6 followed by a 4-ft x 0.085-in. column filled with Carbowax 1540 coated on 35-mesh Teflon 6. The Carbowax column is used to preferentially retard the water peak so that its retention time will fall between those attributed to methanol and ethanol. If a flame ionization detector is used, the Carbowax column may be omitted because water passes undetected through the flame. Although a helium flow rate of 20 ml/min is normally employed when an analysis is being performed, the flow was reduced to 5 ml/min when the GC unit was not in use. However, as noted by Natelson and Stellate, bottles containing the glass beads impregnated with copper sulfate may be prepared in advance, stored, and used as needed. During the development of the chromatogram, the 5-ml chamber (test bottle) is replaced with a fresh one in order to permit the injection of a second sample.

Machata [132] used two chromatographs equipped with flame ionization detectors and, for routine analyses, two columns, one containing 15% PEG 1500 on 60-100 mesh Celite 545 (P-E column "K") and the other 15% PEG 1540 on silanized 80-100 mesh Chromosorb W. In his procedure, a 0.50-ml specimen of urine, blood, or serum is pipetted into a 20-ml serum bottle followed by the addition of 0.10-ml of either a 0.25% acetone or a 0.20% methyl ethyl ketone solution (both used as internal standards). After the introduction of the internal standard, the bottle is securely closed with a serum stopper and then heated to 60°C in a thermostated chamber. A vapor sample is then withdrawn with a syringe previously heated to the same temperature and finally injected into the chromatograph, whose column is maintained at 100°C.

Wallace and Dahl [139], using a Barber-Colman model 20 gas chromatograph, a flame ionization detector, and a column packed with 4.5% of a ternary mixture of liquid stationary phases (Flexol 8N8, diisodecyl phthalate, and PEG 600 in a 1:1:1 ratio) coated on 100-120 mesh Chromosorb P, prepared their samples for GC analysis in the following manner:

A 0.20-ml sample of blood, urine, or other body fluid is placed into a 25-ml Erlenmeyer flask to which is added 0.20 g of sodium chloride. (The salt addition, according to Wallace and Dahl, markedly increases the vapor pressure of ethanol in the biological samples.) After the salt has been added, the flask is capped with a rubber stopper and placed in a 30°C water bath for 5 min. One milliliter of the equilibrated vapor sample is withdrawn with a syringe and injected into the chromatograph.

For this type of analysis, the column was maintained at 75°C with the following regulated gas flow rates: nitrogen carrier gas, 19 ml/min; hydrogen, 22 ml/min; air, 375 ml/min.

Savory and co-workers [142] developed a modified GC procedure of Natelson and Stellate [134,135] which allowed serum ethanol analyses to be performed with increased sensitivity, precision, and convenience. As noted by Savory et al., their modified procedure is similar to the GC method for the determination of lactic acid in blood which was described in 1966 by Savory and Kaplan [155]. The Natelson-Stellate technique is one wherein the sample is dehydrated by addition of anhydrous copper sulfate, and the liberated ethanol is swept by a stream of carrier gas onto the gas chromatographic column.

Serum (50 μl) is injected into a gas tight dehydration tube (Fig. 2.18) which contains 150 mg of powdered copper sulfate (vacuum-dried overnight at 105°C) and is heated by a water-jacketed circulating bath maintained at 80°C. In this heated environment, water is rapidly extracted from the serum by the formation of copper sulfate pentahydrate, an exothermic reaction which simultaneously liberates serum ethanol as a vapor. After a 2-min waiting period, the ethanol vapor is swept by a helium gas stream into a Hallcomid M-18/C-22 firebrick-packed column. Using the chromatographic operating conditions listed in Table 2.13, which also includes the

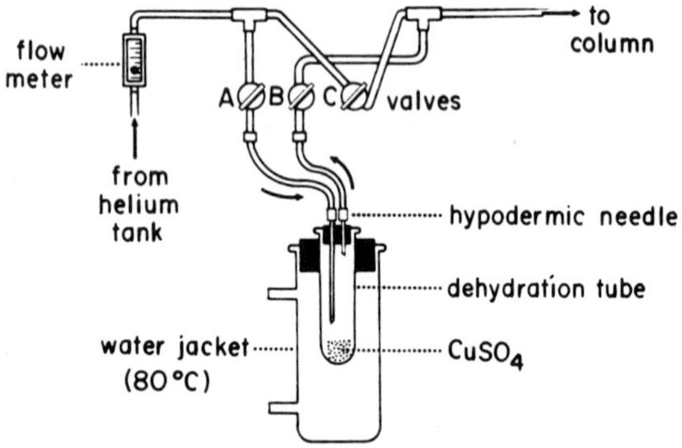

Figure 2.18. Diagram of vapor-injection system for gas chromatographic measurements of serum ethanol. From Savory et al. [142], courtesy of Clinical Chemistry.

TABLE 2.13

Differences Between Savory et al. and Natelson-Stellate Procedures[a]

Parameter	Natelson-Stellate method	Savory et al. method
Type of sample	Whole blood	Serum
Volume of sample	100 μl	50 μl
Dehydrating agent	$CuSO_4$ (250 mg) on perforated glass beads	$CuSO_4$ powder (150 mg)
Heating assembly	Air-jacketed aluminum heating block	Water-jacketed circulating heater
Vaporization temperature	50°C	80°C
Heating time	3 min	2 min
Purging time	0.3 min	1.5 min
Helium flow rate	20 ml/min	75 ml/min
Column packing	Carbowax & Hallcomid on 35-mesh Teflon 6	Diisodecylphthalate, polyethyleneglycol, & Flexol 8N8 on C-22 firebrick
Column temperature	50°C	100°C
Detector	Thermal conductivity	Hydrogen-flame ionization

[a]From Savory et al. [142], courtesy of Clinical Chemistry.

differences between the Savory et al. and Natelson-Stellate procedures, ethanol is eluted from the column in 6.7 min.

With this modified technique, Savory et al. reported a coefficient of variation of 2.0% based on 22 replicate analyses of a single serum sample and 98.7 ± 3.8% recovery for ethanol added to serum. For practical purposes, the limit of sensitivity for detection of ethanol in serum was found to be approximately 0.1 mg/100 ml, which is sufficient for assaying constituents normally found in serum. Furthermore, by decreasing the helium carrier gas flow rate from 75 to 45 ml/min, more complete resolution of an aqueous solution containing a mixture of acetaldehyde, methanol, acetone, ethanol, isopropanol, and n-propanol was obtained, the retention time of ethanol increased from 6.7 to 13.4 minutes with those of the other constituents proportionately prolonged.

Whereas Basette and Glendenning [143] reported their data for a method similar to that of Goldbaum et al. [133], Glendenning and Harvey [146] used sodium fluoride rather than sodium chloride as recommended by Wallace and Dahl [138], and Duritz and Truitt [131] developed a method for ethanol-acetaldehyde analysis in which sample deproteination is accomplished by the addition of zinc sulfate and barium hydroxide prior to sampling the headspace volatiles. Hauck and Terfloth [145] evaluated the errors associated with the determination of blood alcohol by the headspace method using a Multifract F-40. Carried out partly on aqueous solutions of ethanol ranging in concentrations up to 4 mg/ml with t-butanol (2 mg/ml of aqueous solution) as internal standard and partly on blood samples to which the IS was added, they concluded from their study that:

1. An increase of 1°C in bath temperature resulted in a 5.5% increase in the peak heights of ethanol and the standard.
2. The concentration of sodium fluoride influenced individual peak heights as well as the quotient (peak height ethanol/peak height t-butanol); the quotient was reduced by 5.5% when 80 mg/ml of sodium fluoride was added.
3. Repeatability within a week was very good, with a standard deviation of less than ±2%.
4. When blood samples were analyzed, the results had a greater scatter over a larger range than those of aqueous solutions.

Since bacterial production of ethanol from carbohydrate or tissue is possible in postmortem samples, Blackmore [156] suggested a stepwise analytical sequence for specimen selection to be followed for the determination of ethanol as follows:

1. All GC ethanol analyses should be performed with two columns using dissimilar operating conditions.
2. Urine from an intact bladder is preferred for alcohol analysis, since the likelihood of ethanol production by bacterial contamination is rather unlikely.
3. In the event that a good urine specimen is unavailable, blood sampling should be from left and right heart and one other peripheral source. Should the presence of n-butyric or isobutyric acid be

detected (their formation being the consequence of bacterial action which concurrently produces ethanol), the sample must be regarded as suspect.

4. Finally, in both the absence of blood and urine, muscle should be sampled from three peripheral sites and stored at 0°C before homogenization with distilled water and treatment following the procedure normally used for blood specimens.

Concerned with the kinetics of metabolism of small doses (< 140 μg/g) of ethanol when coadministered with trichloroethanol (an alcohol dehydrogenase inhibitor), Gessner [147] required a more sensitive analytical procedure than that of Duritz and Truitt [131] (500 μg/ml maximum sensitivity) and Wallace and Dahl [139] (240 μg/ml maximum sensitivity). Gessner determined ethanol and other aliphatic alcohols in tissue homogenates at low concentration levels by the formation of a more volatile derivative of the agent to be assayed, in this case, the formation of alkyl nitrites by esterification with nitrous acid, which permitted analyses of concentrations down to less than 1 μg/ml. The apparatus used for this study was a Tracor MicroTek model 220 gas chromatograph with a dual hydrogen detector. Using 6-ft x 1/4-in.-o.d. glass columns packed with 120-140 mesh Chromosorb 101 (conditioned for nearly 4 hr), the operating conditions were as follows: nitrogen carrier gas flow rate, 60 ml/min; column temperature, 145°C; inlet temperature, 195°C; detector temperature, 250°C.

Using an assay procedure in which all addition, mixing, and sampling operations are timed from the start of the addition of either the ethanol standard or test sample (the supernate from a tissue homogenate), the GC peak attributed to ethyl nitrite is over 200 times greater than underivatized ethanol and linearly related to the concentration of ethanol in the test sample in the range 0.5 to 10.0 μg/ml.

Sample preparation is rather simple. Into a 35-ml serum bottle, 0.2 g of urea is weighed. Following the urea addition, the bottle is stoppered with a rubber serum cap, purged for 5 min with nitrogen, and then wrapped in aluminum foil. Using tuberculin syringes and with intermittent mixing, 1 ml of the test solution containing ethanol, 0.5 ml of a saturated sodium nitrite solution, and 1 ml of a saturated oxalic acid solution are introduced sequentially into the container. Following a prescribed period of time after the sample was initially introduced into the serum bottle (that is, 3.60 min), a 500-μl sample of the gas phase is withdrawn with a gas syringe; this gas sample is then transferred onto the chromatographic column. At a column temperature of 145°C, the retention times of the pure alcohols and their nitrite derivatives are identical, as can be seen in Table 2.14; in each instance, the sensitivity of the method is sufficiently high to permit detection of less than 1 μg/ml of the alcohol in question. As noted by

Gessner, the sharpness of the peaks as indicated in Figure 2.19 allow separation of a mixture of alcohols and thereby concurrent assay. Furthermore, in duplicate ethanol assays of tissue samples (N = 34), a significantly ($P < 0.005$) greater degree of reproducibility is achieved if methanol (as an internal standard) is added to the sample during homogenization.

Luckey [148] developed a special-purpose gas chromatograph (the Alco-Analyzer), having a constant-volume sample loop, dual columns, and a dual-thermistor detector, for the headspace analysis of ethanol in blood, urine, and breath. In its design, the constant-volume sample loop with its switching valve that permits sample collection, which subsequently can be introduced onto the column by changing the valve's position, is connected to one column (each packed with 80-100 mesh Porapak S) and is located within the column oven, which is maintained at 132°C.

For the determination of blood ethanol, a 1-ml sample is introduced into a 35-ml serum bottle which, after immediately capping with an air-tight rubber injection stopper, is placed into a thermostatically controlled water bath or heating block. When temperature equilibration is reached (usually after about 15 min), a sample of the vapor phase above the blood

TABLE 2.14

Retention Times for Pure Alcohols and for Products of a Reaction Between Alcohol and Nitrous Acid[a] (as determined by sampling of the vapor phase in contact with the reactants)

	Retention time (min)	
Alcohol	Pure alcohol	Reaction product
Methanol	1.3	1.3
Ethanol	2.0	2.0
n-Propanol	3.9	3.9
Isopropanol	2.8	2.8
n-Butanol	8.2	8.2
sec-Butanol	6.3	6.3
tert-Butanol	3.8	3.8
Isobutanol	6.6	6.6

[a]From Gessner [147], courtesy of Analytical Biochemistry.

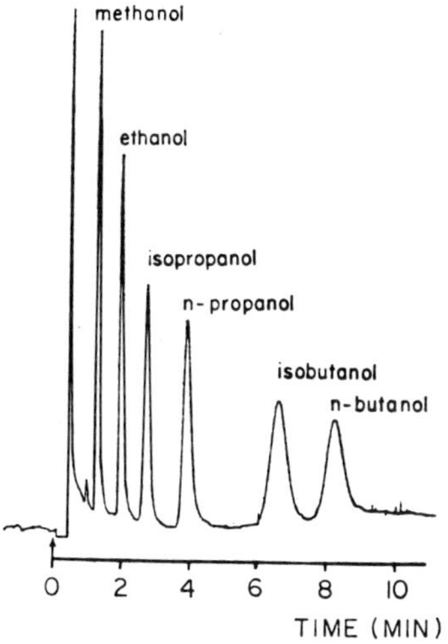

Figure 2.19. Gas chromatogram of 500 μl of vapor phase in contact with reaction mixture of 1.5 ml of nitrous acid and 1 ml containing 10 μg each of methanol, ethanol, n-propanol, isopropanol, n-butanol, and isobutanol. Conditions: 6-ft glass columns, 1/4-in. diameter, packed with Chromosorb 101, 120-140 mesh; column temperature 145°C; nitrogen flow 60 ml/min; flame ionization detector; attenuation 10 x 2. From Gessner [147], courtesy of Analytical Biochemistry.

specimen is withdrawn with a syringe and injected into the chromatograph where the GC loop isolates a fixed volume of this transferred headspace gas. This isolated sample is then introduced onto the column by selecting the proper position of the switching valve.

Coldwell et al. [149] prepared the blood specimen by taking a known volume up in a disposable pipette rinsed with heparin and then transferring this sample to an air-dried filter paper disk (impregnated with a 0.2% sodium fluoride solution) contained in a 25-ml serum bottle. After the tip of the disposable pipette is broken off and also added to the bottle's contents, the container is capped and equilibrated. As in other headspace techniques, an aliquot of gas is removed and injected into the gas chromatograph. In their study, the GC instrument was equipped with flame ionization detectors and a column filled with 5% Carbowax 1500 coated on 30-60 mesh Haloport 60F. With column, injector, and detector temperatures of 75 to 100°C, 130°C, and 200°C, respectively, the various gases were regulated to provide the following flow rates: nitrogen carrier gas, 30 ml/

min; air, 1000 ml/min; hydrogen, 30 ml/min. For quantitative determinations, calibration curves were prepared using standard solutions handled in the same manner as blood specimens.

In 1972, Machata [150] described the experimental and theoretical bases for the GC determination of blood alcohol for both routine and automatic analysis. For the analysis of the vapor over the solution at a fixed equilibrium temperature using t-butanol as an internal standard (its vapor pressure paralleling that of ethanol in the 45 to 70°C temperature range so that the ratio of the vapor pressures in this range is independent of the temperature as indicated in Fig. 2.20), Machata employed the Perkin-Elmer Multifract F-40, which permitted the unattended automatic analysis of a large number of samples and headspace analysis under the most diverse operating conditions. Although Carbowax 1500 or 1540, Porapak, and Hallcomid 18 columns are acceptable stationary phases as illustrated in Table 2.15 by the retention times of toxicological substances on these three columns, the following gas chromatographic conditions were selected as being the best for blood alcohol determination:

Column	2-m packed with 15% Carbowax 1500 on Celite
Column temperature	70°C
Carrier gas	Nitrogen
Inlet pressure	1 atm (14.7 psig)
Flow rate	25 ml/min
Attenuation	64 x (2.5-mV recorder)

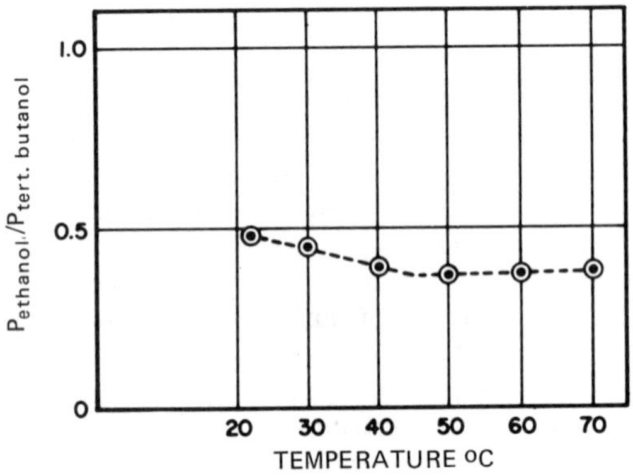

Figure 2.20. The ratio of the vapor pressures of ethanol and tert-butanol in the vapor phase over a 0.1% solution, as a function of the temperature. From Machata [150], courtesy of Perkin-Elmer Corporation.

TABLE 2.15

Retention Times of Various Substances Observed on Three Columns[a]

Stationary phase	Carbowax 1500	Hallcomid 18	Porapak
Column temperature (°C)	90	90	175
		Retention time (min)	
Methanol	2.1	2.6	1.1
Ethanol	2.6	4.0	2.1
Isopropanol	2.5	4.8	3.4
n-Propanol	4.2	8.7	4.5
n-Butanol	7.2	19.3	10.2
Acetaldehyde	1.1	1.7	1.5
Diethyl ether	0.8	1.3	4.2
Acetone	1.5	2.1	3.2
Chloroform	2.8	2.4	7.2
Paraldehyde	4.6	12.7	4.3
Methyl ethyl ketone	2.1	4.2	6.9
tert-Butanol	2.2	5.4	5.4

[a]From Machata [150], courtesy of Perkin-Elmer Corporation.

The Perkin-Elmer Multifract F-40 consists essentially of three building blocks:

1. A standard compact gas chromatograph with a flame ionization detector and equipped with a special dosing system besides the conventional injection port

2. A turntable accommodating 30 samples whose temperature is precisely controlled by a water thermostat

3. The necessary control unit adjusting the variable conditions and indicating by signal lights the individual stages of the automatic analysis

After the rotor delivers the sample bottle to the instrument, its headspace gas is sampled and the gaseous sample delivered into the gas chromatograph with the help of a special electropneumatic dosing system for analysis. After the analysis is finished, the turntable disconnects the sample bottle from the dosing system and delivers a new sample. If desired, each analysis can be repeated.

Standard high-form serum bottles of 20-ml capacity are used as sample bottles. A volume of 0.5 ml of the specimen to be analyzed is pipetted into the bottle, 0.1 ml of the internal standard (a 0.2% w/v solution of t-butanol in water) is added, the bottles are closed with a standard rubber septum, and the covers are secured with aluminum caps for the eventuality of an excess pressure buildup. The sample bottles are then placed into the turntable, which is thermostatted at 60°C.

The electropneumatic dosing system consists of a T-shaped line; one end is connected to the front of the separation column, the second end leads to the dosing needle, and the third end is connected to the carrier gas line by a magnetically operated valve. The entire dosing system is thermostatted. Its operation can be described as follows (see Fig. 2.21):

At the beginning (step 1) the dosing needle is closed by a mobile cylinder having a rubber septum at its lower end. Now the turntable positions the sample bottle under the dosing system and lifts it up. The sample bottle will shift the mobile cylinder up so that the dosing needle pierces the rubber septum of the cylinder as well as the rubber cap of the sample bottle, and penetrates into the bottle (step 2). This step also provides a connection with the carrier gas line through the magnetic valve M--which is open--and thus, the pressure in the sample bottle is built up to the pressure of the carrier gas before the column. Then, the magnetic valve closes cutting off the carrier gas supply. Now, there is a connection only between the gas space of the sample bottle and the column, and the pressure is reduced by a gas flow from the sample bottle through the needle to the column. While the gas passes from the bottle to the column, a definite volume of the sample headspace is flushed into the column (step 3). The dosing

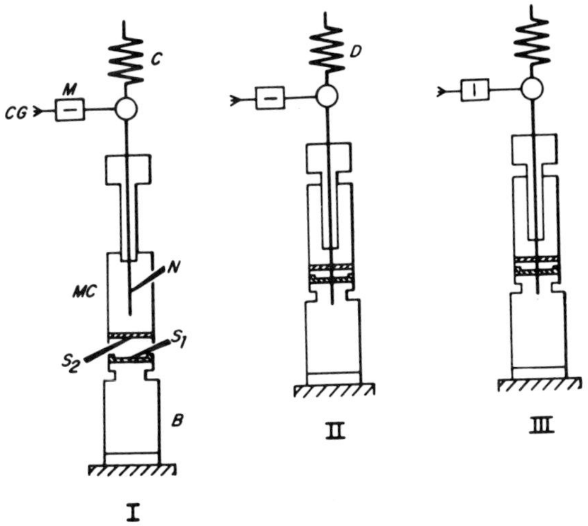

Figure 2.21. The schematic of the dosing system and its operation. C, column; M, magnet valve; CG, carrier gas; N, dosing needle; MC, mobile cylinder; B, sample bottle; S, septum closing the sample bottle; S_2, septum closing the mobile cylinder. From Machata [150], courtesy of Perkin-Elmer Corporation.

process is completed when the magnetic valve opens the carrier gas supply again; then the dosing needle is flushed by the carrier gas, the sample bottle is removed, and a new bottle is delivered by the turntable or, if desired, the analysis is repeated.

The actual dosed amount (volume) of each gas sample depends on the column inlet pressure, the gas volume of the bottle, and the time interval during which the magnet valve is closed. In the routine determination of blood alcohol, the column inlet pressure is at 1 atm (gauge, 14.7 psig), the gas volume is kept constant by the use of constant volumes of blood samples and standardized bottles, and the sampling time was chosen as 2 sec; under these conditions the gas volume introduced into the column is about 0.5 ml. The time for pressure buildup is 1 min, and the time to flush out the dosing needle is 12 sec. For blood alcohol determination, the dosing system is thermostatted at 125°C. The samples are changed every 4 min, and each analysis is carried out in duplicate.

At the start of each analysis the control mechanism releases an impulse which, in accordance with the sample number of the bottle, produces a deflection on the recorder which is proportional to the sample number. An optional electronic unit simultaneously adjusts the zero line on the recorder.

The turntable is equipped with adjustable contacts which automatically turn the system off after any preset number of samples. This is particularly important when the analysis of a series of samples is carried out overnight.

The instrument is also equipped with a standard injection block which permits manual sample injection and use of the gas chromatograph in the Multifract F-40 without the automatic dosing system. In this case the connection of the separation column has to be changed [150].

For quantitative evaluation of the chromatograms, the ratio of the heights or areas of the ethanol and t-butanol peaks is used, the height measurements being preferred since it is as accurate as area measurement by integrator (see Table 2.16).

The ethanol concentration in the sample is calculated using standard ethanol samples of known concentration and the amount of internal standard added to the samples. Having calculated a calibration factor f from standard samples run parallel to the actual samples from the expression

$$f = C_s \left(\frac{H_b}{H_e}\right) \qquad (2.20)$$

where C_s is the ethanol concentration in the standard sample and H_b and H_e are the respective heights of the t-butanol and ethanol peaks obtained when analyzing the sample, the ethanol content in the unknown specimen is given by

$$\text{Ethanol } \% = \left(\frac{H'_e}{H'_b}\right)\left(\frac{f}{F}\right) \qquad (2.21)$$

where H'_e and H'_b are the respective heights of the ethanol and t-butanol peaks, f is the calibration factor, and F is a correction factor (for blood samples F = 1.08, and for serum F = 1.26).

In a continuous test in which 30 samples were analyzed, the standard deviation was computed as ±0.63%, resulting in a spread of ±1.89% (equivalent to 3σ). The reproducibility of the sampling process (evaluated by investigating the variations of the peak heights in repetitive analyses) yielded a standard deviation of 1.1%.

An interesting article is that by Vitenberg et al. [153], who discussed and compared the different ways of GC determination of trace impurities using phase equilibria of the objects under investigation (headspace analysis and the determination of impurities in a gas using the data of their content in the liquid saturated with the gas). Resorting to relationships that exist between the partition coefficient (K_T) of the compound of interest and the equilibrium concentrations of the compound in the liquid (C_L) and gas (C_G) at temperature T, Vitenberg et al. show that the initial concentration of the component of interest in solution, $C_L{}^0$, is given by

TABLE 2.16

The Ratio of the Area and Height of the Peaks Corresponding to the Substance of Interest and the Internal Standard (tert-Butanol), Obtained on the Carbowax 1500 Column at 90°C

Substance	Retention time (min)	Ratio of the peak areas	Ratio of the peak heights
Diethyl ether	0.8	9.15	10.44
Acetaldehyde	1.1	2.26	2.36
Acetate	1.5	0.52	0.70
tert-Butanol[b]	2.2	-	-
Ethanol	2.6	0.51	0.47
n-Propanol	4.2	0.48	0.46
Paraldehyde	4.6	1.19	0.56
n-Butanol	7.2	1.16	0.37

[a]From Machata [150], courtesy of Perkin-Elmer Corporation.
[b]Internal standard.

$$C_L^0 = \frac{C_G(K_T V_L + V_G)}{V_L} \tag{2.22}$$

where

V_L = liquid phase volume
V_G = gas phase volume

Although one must know the values for the various constants shown in Eq. (2.22) to determine C_L^0, precise measurements of the two phase volumes can be avoided by performing a calibration with a standard solution under standard conditions.

In the situation where K_T is independent of compound concentration, for example with dilute solutions where the activity coefficients of the compounds being studied are essentially constant or nearly equal to unity in the concentration range of interest, they note that, in accordance with Henry's law,

$$C_L = \frac{P'_x}{\gamma P_x^0} \tag{2.23}$$

where

P'_x = partial pressure of component x above the solution at temperature T
P_x^0 = the vapor pressure of pure component x at temperature T
γ = activity coefficient

Thus, in the headspace method of analysis, the component's peak height or peak area determines the magnitude of P'_x. Provided that the product γP_x^0 for the compound under study is known, it is possible to calculate C_L. Furthermore, from Eq. (2.23), in the case of nearly ideal or ideal diluted mixtures the component's peak area or height depends linearly on its concentration in solution. It follows, therefore, that C_L^0 can be obtained from Eq. (2.22). However, as stressed by Vitenberg et al., the peak height of the component for real mixtures is nonlinearly dependent on C_L and, under these circumstances, an accurate preliminary calibration is required for quantitative analysis of admixtures in solutions, the calibration carried throughout the concentration range covered.

Since K_T is temperature dependent, the need for a constant temperature is obvious; the accuracy of thermostating in the course of equilibration and sampling is determined by the shape of the partition coefficient-temperature dependence and the allowable error of analysis.

In addition to the above, the limits of application and the possible source of error for the method are discussed and examples for the application of the method of equilibrium concentration for GC analysis of impurities in food, waste water, air, and biological media are given.

B. Drug Treatment of Chronic Alcoholism

As noted previously, the metabolic pathway for normal ethanol oxidation can be altered chemically by disulfiram, metronidazole, or an antidiabetic sulfonylurea such as tolbutamide (the analysis of this latter, rarely used drug will be discussed in a subsequent volume).

To date, of the several metabolic products of disulfiram reported in the literature (for example, sulfate [157], carbon disulfide [157-160], hydrogen sulfide [158], diethyldithiocarbamic acid [157-159,161-164], methyl ester of diethyldithiocarbamic acid [164], and N,N-diethyldithiocarbamoyl-1-thio-β-D-glucopyranosiduronic acid [157,159,165]), only carbon disulfide has been detected in expired air by gas chromatography [166]. This last metabolite is formed by the metabolic pathway suggested for disulfiram by Gessner and Jakobowski [164]:

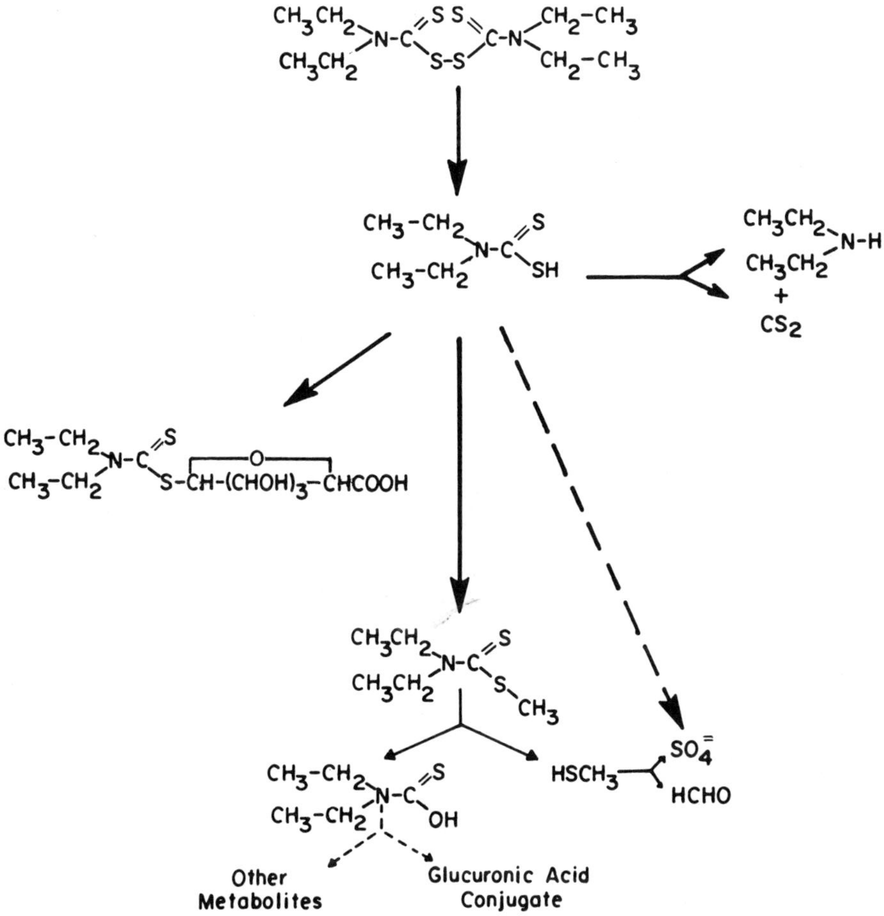

Other
Metabolites
Glucuronic Acid
Conjugate

Wells and Koves [166] described a method for CS_2 detection in breath in which expired air is bubbled through 3 ml of a 5% solution (v/v) of diethylamine in acetone contained in a 30-ml Drechsel-type glass bubbler. After five exhalations, the solution is decanted into a 15-ml centrifuge tube, 6 drops of methyl iodide are added, and finally its volume is reduced almost to dryness at 50°C with a nitrogen stream. To this reduced volume, the following are added in sequential order: (1) 2-ml of distilled water, (2) 3 drops of 3 N sulfuric acid, and (3) about 1 ml of n-butyl chloride. The resulting mixture is shaken on a Vortex mixer for 1 min and then centrifuged. Following the transfer by pipette of the upper organic layer into a 3-ml Reacti-vial, the solvent is again evaporated with nitrogen at 50°C. After the addition of 50 μl of n-butyl chloride to the residue, 5 μl is withdrawn and injected into the gas chromatograph. The GC unit is equipped with a flame ionization detector and a 4-ft x 1/4-in. coiled glass column packed with 1.5% FFAP on 80-100 mesh Varoport 30. With the column, injection, and detector temperatures maintained at 125, 200, and 220°C, respectively, and the nitrogen carrier gas flow rate set at nearly 40 ml/min, the methyl ester of diethyldithiocarbamic acid is eluted in approximately 5.2 min. Detected in submicrogram quantities with a flame detector, the CS_2 in expired air is analyzed as the ester which is formed as shown by the reactions given below:

$$2\,Et_2NH + CS_2 \longrightarrow Et_2NCS_2^- + \overset{+}{N}H_2Et_2 \qquad (2.24)$$

$$Et_2NCS_2^- + CH_3I \longrightarrow Et_2NCS_2CH_3 + I^- \qquad (2.25)$$

Although metronidazole (Flagyl) is an effective drug in the treatment of Trichomonas infections as well as an effective amebicide, it may elicit an "Antabuse reaction" when alcohol is ingested [1]. Yet despite several claims of its "disulfiram-like effect", Gelder and Edwards [167] have reported that it has not been found to be any more useful than a placebo in treating alcohol addicts [168].

Analytically, however, it can be determined at therapeutic levels in plasma by gas chromatography. Midha, McGilveray, and Cooper [169] developed a quantitative procedure for metronidazole (1-β-hydroxyethyl-2-methyl-5-nitroimidazole) which permitted concentrations of 0.25 μg/ml of plasma to be determined with a precision of ±6%. Applicable to routine estimation of plasma levels following single therapeutic doses, the proposed procedure for metronidazole analysis consists of adding 1 ml of plasma to 1 ml of an aqueous solution of the internal standard, 1-β-hydroxyethyl-2-methoxy-5-nitroimidazole (15 μg/ml), which has previously been pipetted with 4.0 ml of 0.1 M carbonate-bicarbonate solution into a 42-ml centrifuge tube. Following the addition of 15 ml of a binary solvent extraction mixture (methylene chloride-diethyl ether), the contents are shaken for 30 min on a flat Eberbach shaker (216 cycles/min). By pipette,

13 ml of the upper organic phase is transferred into a conical glass-stoppered tube (45 ml). Having repeated the extraction twice with two 15-ml portions of the solvent mixture (these extracts also added to the conical tube), the combined extracts are evaporated to nearly 1 ml using a 60°C water bath and a dry stream of nitrogen. The residue is dried with anhydrous calcium chloride, transferred quantitatively with a pasteur pipette, using two 0.5-ml portions of methylene chloride, into a 3-ml glass-stoppered conical tube and then redried in the same manner as noted above. To the residue, 50 and 30 μl of acetonitrile and Tri-Sil concentrate, respectively, are added in order to prepare trimethylsilyl (TMSi) derivatives; this is accomplished overnight at room temperature. Again the sample was evaporated at room temperature and redissolved with 50 μl of acetonitrile. From this final solution, 1 to 2 μl arc withdrawn and injected into a Perkin-Elmer F-11 gas chromatograph equipped with a flame ionization detector and a 6-ft x 1.5-mm-i.d. spiral glass column packed with a mixed liquid phase (1% Apiezon L, 0.5% SE-30, and 0.5% OV-17) on 80-100 mesh Chromosorb W. Following column silanization and conditioning, Midha et al. selected the following operating conditions for metronidazole analysis: column temperature, 135°C; injector temperature, 210°C; detector temperature, 210°C; nitrogen carrier gas flow rate, 80 ml/min.

Using the conditions described, the TMSi derivatives of the drug and the internal standard were approximately 4.9 and 12.0 min, respectively. The structures of the TMSi ethers were confirmed when these materials were analyzed with a GC-MS integrated instrument (a Hitachi Perkin-Elmer model RMS4 mass spectrometer interfaced via a two-stage jet separator to a P-E gas chromatograph model 990).

With this procedure, it was possible to monitor plasma levels measured at various time intervals in a man following the administration of a single 250-mg oral dose of metronidazole.

III. VOLATILE CONSTITUENTS IN HUMAN BREATH, FLUIDS, AND TISSUES

During the past decade, there has been an ever-increasing research effort made by many investigators to develop mass screening procedures for the diagnosis of disease based on the gas chromatographic analysis of volatile constituents in biological fluids. This is an area of great importance since, as stressed by Robinson et al. [170]:

> It has been suggested that the treatment of disease should be carried out by adjustment of the concentrations of substances in the human

body so as to duplicate those concentrations which correspond to optimum health and that this adjustment should be carried out whenever possible by the administration of compounds normally found in the human body [171]. The administration of the right molecules in the right amounts is called "orthomolecular therapy." It has been further suggested that a quantitative assessment of the state of molecular health of the human body, "orthomolecular diagnosis," might be made in a practical way by the quantitative analysis of the normal constituents of human urine followed by a simple type of computerized pattern recognition [172].

In addition to the above, GC and MS are also excellent analytical tools for determining low-molecular-weight drugs and other toxicological materials as well as metabolites of more complex substances that have been converted to more volatile by-products. In the analysis of various body fluids and tissues for their ethanol content, one has already been exposed to some of the volatile constituents that have been observed. Not restricted solely to low-boiling, highly volatile compounds, high-molecular-weight substances have been separated and identified chromatographically for similar purposes: the determination of "metabolic profiles" for each individual.

A. Breath Analysis [63,65,148,173-189]

After ingestion and absorption, the alcohol in blood is in equilibrium with alcohol in alveolar air at a given temperature. As noted by Jain and Harvey [63] in their excellent review of breath alcohol methods, its breath concentration is controlled by a diffusion process, is governed by Henry's law, and is a predictable phenomenon since the body temperature is essentially constant. For example, studies [190-192] have shown that about 2100 ml of alveolar air (when measured at 34°C, the average temperature at which the breath leaves the mouth) contain the same amount of alcohol as 1 ml of blood provided that the interval between drinking and sampling is long enough to avoid contamination of the alveolar air by fluid alcohol. Spector [193] states that breath alcohol tests are totally unreliable for at least 20 min after alcohol exposure unless large correction factors are applied to readouts of breath analyzer instruments. Furthermore, as pointed out by Payne [194] and by Milner and Landauer [195], some inaccuracies encountered by breath methods result from attempts to correlate factors that are not strictly comparable, such as pulmonary blood and peripheral venous blood.

As noted in their article, Jain and Cravey also discuss other methodologies for alcohol analysis in addition to those performed by gas chromatography: namely, chemical [196-201], electrochemical [202,203], conductance [204], heat of oxidation [205], infrared absorption [206], and dehydrogenation [207,208] methods.

Several GC instruments have been developed for alcohol analysis (the Alco-Analyzer by Luckey [148] and the Intoximeter [200], which are both capable also of determining blood and urine alcohol levels).

Whereas the Alco-Analyzer has been previously discussed as a headspace gas analyzer, the Gas Chromatographic Intoximeter (GCI) was designed specifically to handle direct or field-collected breath, urine, and blood alcohol specimens. Its design is such that it incorporates a multisampling valve, injector, column, and flame ionization detector in a single chassis. With alcohol concentrations reported automatically in 90 sec with both an internal strip-chart recorder and a digital readout, the GC unit is equipped with a 12-in.-long column filled with Porapak Q and a single-source mixture of nitrogen carrier (60%) and hydrogen flame fuel (40%) gases. Performing analyses on 1/4-ml deep lung air samples, a sample loop is used to collect the direct breath sample after the upper lung or tidal air is vented into a waste bag of fixed volume. Field-breath samples are collected in an indium tube which provides three hermetically sealed capsules of breath, a technique used by Morris et al. [188], who found excellent correlation between GCI data and blood alcohol tests performed by a distillation-dichromate titration method.

In addition to these instruments, which were designed specifically for alcohol analysis, trace compounds in body fluids which might prove valuable in diagnosis or nutritional studies have been determined at parts-per-billion levels by gas chromatography [173]. In 1960, Thompson [173] envisioned a broader applicability for gas chromatography as a research tool to study not only the physiology and various aspects related to manned space flight, but human body chemistry itself for the detection of disease and illness, characterizing human physical capabilities, and correlating human endurance and efficiency with such things as nutritional deficiencies and trace constituents in the environment. Preliminary studies conducted at that time by the Southwest Research Institute tentatively identified by retention time comparisons of known-unknown samples 15 compounds (Table 2.17) present in parts-per-million or parts-per-billion levels in urine and expired air.

In 1969, Jansson and Larsson [174] analyzed organic compounds in human breath by gas chromatography-mass spectrometry and investigated sampling methods. By use of a respiratory mask, samples were collected in plastic Mylar bags, which could be stored for a considerable length of time before analysis. Whereas quantitative analyses were made by direct

TABLE 2.17

Volatiles in Breath and Biological Fluids[a]

Compounds[b]	Expired air	Urine	Blood
Acetone	x	x	x
Acetaldehyde	x	x	x
Isoprene	x	x	x
Methanol	x	x	x
Ethanol	x	x	x
Methyl furan	x		x
Furan	x		
Methyl formate		x	
Propionaldehyde	x	x	
Methyl ethyl ketone		x	
Diacetyl		x	
Dimethyl sulfide	x	x	
Diethyl sulfide	x		
Isovaleraldehyde	x		
Acetonitrile		x	
Unidentified	4	3	2

[a]From Thompson and McKee [173], courtesy of Chemical and Engineering News.

[b]Identification is tentative.

injection of 5-ml breath samples into gas chromatographs equipped with flame ionization detectors, compound identification was performed after sample enrichment with an integrated GC-MS system (LKB 9000).

For methane analysis, a Linde 5A molecular sieve column was used at 70°C while the other organic compounds (Table 2.18) were separated on either a Porapak Q column at 110°C or a 20% PEG 1500 at 65°C.

For identification purposes, sample enrichment was made with small glass tubes containing activated charcoal which were then attached to the gas sampling valve of the LKB instrument. With carrier gas flowing through the tube, the tube was heated for 2 min to 180°C with an electric

TABLE 2.18

Analysis of Breath Volatiles by GC-MS[a]

Component	Subjects tested	Concentration (ppm)	
		Mean	Range
Methane	43	5.1	0-32
Isoprene	13	0.23	0.09-0.45
Acetone	13	0.52	0.22-0.87
Methanol	-	1.05	0.27-3.39
Ethanol	-	0.13	0.05-0.36

[a]Adapted from Jansson and Larsson [174].

oven to free the breath constituents adhering to the surface of the activated charcoal. Compounds and their concentrations measured in healthy subjects with this analytical scheme are listed in Table 2.18.

Trotter, Sulway, and Trotter [184] described a specific GC method for the determination of acetone in breath and plasma. This method of breath sampling was adapted for clinical use at the bedside and suitable for studies on diabetic patients with hyperketonemia.

The GC operating parameters used with a Pye 104 gas chromatograph equipped with hydrogen flame ionization detector and a 1.5-m x 4-mm-i.d. glass column packed with 80-100 mesh Chromosorb 102 were the following: column temperature, 115°C; nitrogen carrier gas flow rate, 50 ml/min. Using these conditions, water, acetaldehyde, ethyl alcohol, acetone, and isopropyl alcohol were eluted from the column in 0.7, 2.0, 4.4, 7.3, and 9.0 min, respectively. Samples were collected in a 1-liter nylon-polyester bag using a breath-sampling procedure based on the rebreathing method of sampling respired gases as originally developed by Collier [209] and Levey et al. [210].

On the other hand, following centrifugation of venous blood collected in lithium heparin-coated tubes, 1-μl of plasma was injected on the column. For 145 duplicate pairs of plasma acetone determinations performed on diabetic patients with hyperketonemia, the coefficient of variation was 5.2% over a plasma acetone range of 0.05 to 12.90 mmoles/liter. A similar analysis of breath acetone determinations showed a coefficient of 5.5% over a range of breath acetone levels of 0.03 to 27.4 μmoles/liter. Although injections of solutions of propane-1,2-diol, lactic acid, pyruvic acid, glucose (200 mg/100 ml) and 3-hydroxybutyrate (10 mmoles/liter) did not yield a chromatographic peak, approximately 10% of the acetoacetate present in the plasma specimen under analysis was thermally degraded on the column (115°C), which contributed to the measured acetone levels.

Using both techniques on patients with uncontrolled diabetes [211,212], breath and plasma acetone were shown to be linearly related in nine cases of diabetic ketoacidosis (coefficient of correlation = 0.928), providing evidence for the validity of the breath-collection method.

Teranishi and co-workers [185] developed in 1972 a more sophisticated technique of direct gas chromatographic analysis of volatiles from human breath and urine, using GC as a diagnostic tool for uncovering sickness in human beings, whether it be from malfunctions of organs and tissues or from invasions of viruses or microorganisms, which would be reflected in quantitative and qualitative changes in the metabolites emitted.

Using dual flame ionization detectors [213], dual large-bore open-tubular columns [1000 ft x 0.03-in.-i.d., 0.0625-in.-o.d. SS tubing coated with SF 96(50) methylsilicone oil] [214] in conjunction with a Varian Aerograph model 2100 gas chromatograph, capable of column temperature programming from 20 to 180°C at a rate of 0.5°C/min, the inlet, valve system, and detector were maintained at 100, 100, and 150°C, respectively, whereas the helium, hydrogen, nitrogen, and air flows (regulated with restrictors and surge tanks) were 15, 50, 50, and 600 ml/min, respectively.

Human breath volatiles from 10 to 15 exhalations per sample were first trapped in a coiled stainless tubing (see Fig. 2.22) immersed in an isopropyl alcohol-Dry Ice bath but, because of the large amount of water in breath

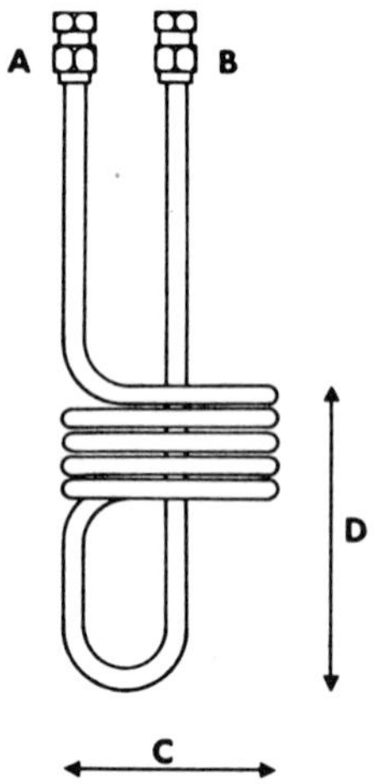

Figure 2.22. Breath trap, made from 0.25-in.-o.d., 0.20-in.-i.d. stainless steel tubing, 5 ft long. A. Breath inlet side. B. Breath outlet side. C. 2.5-in. D. 3.5-in. From Teranishi et al. [185], courtesy of Analytical Chemistry.

and urine samples, a larger trap was constructed as illustrated in Figure 2.23. As in techniques used in flavor research [215], the volatiles from the larger coil were subsequently transferred to the GC system for analysis.

On the other hand, a special urine volatiles sample holder (Fig. 2.24) was constructed which permitted the volatiles to be collected over a 1-hr period in the "double helix" trap. The transfer of volatiles from sample to trap was achieved by passing helium at 15 ml/min over a 200-ml sample which was magnetically stirred and thermostated at 80°C. As noted by the authors, the use of metal columns was a compromise until glass-column technology is improved, recognizing that Horning, Vanden Heuvel, and Creech [216], and Horning [217], had shown the usefulness of glass-packed columns whereas Novotny and Zlatkis [218] explored the possibilities of glass capillary columns for biochemical research studies as well as routine analyses.

Describing modifications of earlier work [185,219] which are more amenable for routine analyses, Zlatkis et al. [186] developed a new procedure for sampling, transfer, and analysis of volatile organic compounds by gas chromatography. In their procedure, an adsorbent glass tube trap (10-cm x 8-mm i.d.) filled with 35-60 mesh Tenax GC is essentially

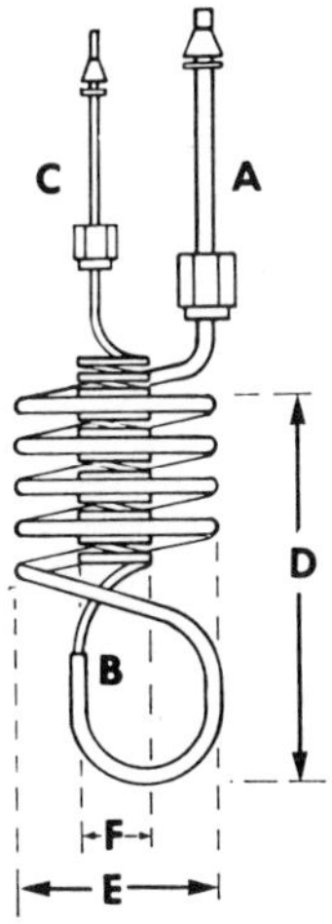

Figure 2.23. "Double-helix" sample trap in gas chromatograph. A. Inlet side, 0.125-in.-o.d., 0.10-in.-i.d. stainless steel tubing, wound into helix. B. Silver solder joint from 0.125-in.-o.d. tubing to 0.0625-in.-o.d. tubing. C. Outlet side, 0.0625-in.-o.d., 0.040-in.-i.d. stainless steel tubing, 10-in. long, wound into helix. D. 3-in. E. 1.5-in. F. 0.5-in. From Teranishi et al. [185], courtesy of Analytical Chemistry.

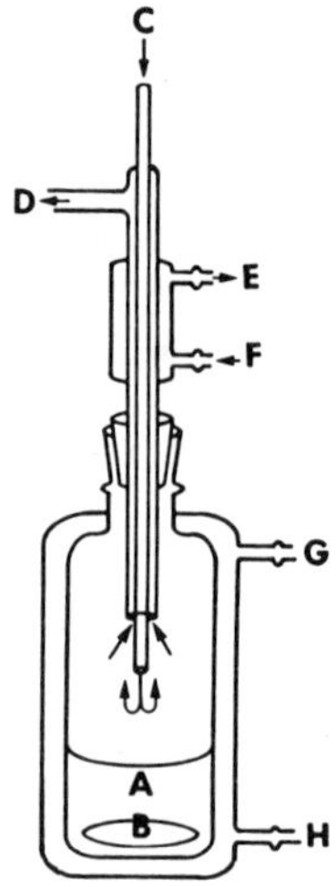

Figure 2.24. Urine volatiles sample holder. A. Urine sample. B. Bar magnet. C. Helium gas in. D. Helium stream out, with urine volatiles. E. Condenser water out. F. Condenser water in. G. Heated water out. H. Heated water into jacket of sample holder. From Teranishi et al. [185], courtesy of Analytical Chemistry.

attached to D in Figure 2.24 with a Teflon sleeve. This is sufficient to adsorb the volatiles released from a 100- to 200-ml urine sample that is heated to 100°C, stirred, and purged with a helium stream adjusted to 20 ml/min. For collecting air or exhalation samples, the Tenax trap is attached to the condenser with the sleeve. In this arrangement, the samples are drawn through the system with vacuum (0.5 to 2.0 liters/min).

After the sample has been purged for 1 hr (prescribed time period for aqueous or urine samples), the Tenax trap is inserted into a modified injector port which fits into a Perkin-Elmer 900 gas chromatograph. With injector and detector temperatures of 300 and 280°C, respectively, the sample is transferred from the adsorbent at 300°C for 20 min at 20 ml/min to a 10-ft x 0.04-in.-i.d. open-tubular precolumn attached to the outlet of the injector port and cooled with Dry Ice (Fig. 2.25). Chromatography is begun when the sample is transferred via the four-port valve to the analytical column (300 ft x 0.02-in. i.d., nickel coated with Emulphor-O). As indicated in Figure 2.25, the four-way valve permits the simultaneous flow of carrier and purging gases. With this technique, Zlatkis et al. studied various subjects' urine and breath for the specific detection of volatile organic pollutants and/or their metabolites and the possibility of making a direct correlation with a specific pollutant profile of a given locale.

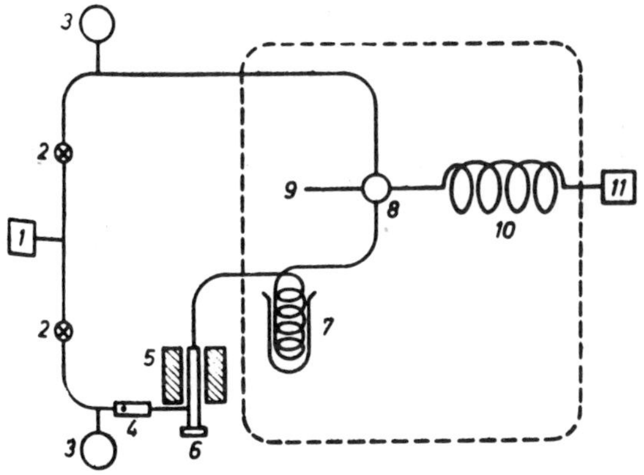

1 Carrier (He)
2 Toggle switch
3 Pressure regulator
4 Rotameter
5 Injector block
6 Modified injector port
7 Precolumn in Dry Ice
8 Four-port valve
9 Vent
10 Separating column

Figure 2.25. Sample analysis and transfer flow system. From Zlatkis et al. [186], courtesy of Chromatographia.

B. Urine Analysis [6-8,85-87,92,95,98-100,107,111,112,116,117, 120,133-135,139,148,170,173,185,186,219-229]

In previous sections of this chapter dealing with ethanol analysis and describing the detection of breath volatiles, many of the procedures have been applied to urine samples, the number of volatiles detected being limited in number and restricted to rather low-molecular-weight, low-boiling components. Similar studies have been also performed by other investigators in addition to human urinary metabolic profiles of more complex structures. Those not previously discussed will be covered in this portion of the text.

In 1962, Bonnichsen and Linturi [99] injected 2 μl of unaltered urine specimens directly into a gas chromatograph with a flame ionization detector for the determination of ethanol levels (see direct injection methods for GC details). Besides ethanol, the analysis revealed the presence of methanol, acetone, and acetaldehyde, but in no instances were aliphatic alcohols higher than ethanol found. As expected, acetone was commonly encountered in diabetic individuals, its concentration being around 0.01%. Although the enzymatic determinations of ethanol agreed rather well with GC data, discrepancies were noted between GC-Widmark results as shown in Table 2.19.

TABLE 2.19

Comparison of Urine Ethanol Data, % (g/L)[a]

Urine		Gas chromatography of the urine		
Widmark (%)	ADH (%)	Ethanol (%)	Methanol (%)	Acetone (%)
3.69	2.47	2.40	0.78	0.02
4.64	4.25	4.29	0.39	0.02
2.39	1.10	0.98	0.66	0.11
3.23	2.54	2.55	0.39	0.04
3.89	3.93	3.74	-	0.20
3.61	2.63	2.73	0.69	0.04
3.29	2.35	2.46	0.45	0.01
1.65	1.12	1.07	0.06	0.17

[a]Adapted from Bonnichsen and Linturi [99], courtesy of Acta Chemica Scandinavica.

Although methanethiol had been detected in human urine after asparagus ingestion, Field et al. [220] were able to show by GC analysis that this component is also present in varying degrees when a subject is placed on a nonasparagus-containing diet. Using both silicone oil and tricresyl phosphate columns in conjunction with a flame ionization-equipped gas chromatograph, the sample consisted of the headspace vapor above and in equilibrium with urine heated to 60°C in a gastight container. For identification purposes, they noted that one of the component peaks (varying with asparagus ingestion) recorded had identical retention characteristics as authentic methanethiol on both columns; the peak was also reinforced by the addition of authentic methanethiol to the urine sample.

Of equal importance was the study performed by Van Haaften and Sie [221], who measured by gas chromatography the phenol concentration in urines taken from individuals exposed to benzene vapor. Comparing GC data with results derived by colorimetric methods on urine steam distillates [230,231], the differences noted were attributable to the presence of cresols in urine which also reacted with the color-producing reagents. However, for a number of adults who were neither exposed to benzene nor on medication whose average urine phenol level was 5 mg/liter, their GC p- and m-cresol average was 89 mg/liter.

As proposed, each urine sample was mixed with an equal volume of concentrated phosphoric acid and, from this acid mixture, a 10-μl sample was injected into a hydrolysis precolumn via the injection port. The precolumns or hydrolysis sections where phenols are liberated from conjugated chemical structures consist of 10-cm lengths of stainless steel tubing (6 mm i.d.) filled with 30-50 mesh borosilicate glass powder. In turn, the nonconjugated phenols are carried to and separated on an analytical column (100-cm x 3-mm-i.d. stainless-steel tubing) packed with 10% (w/w) PEG 6000 coated on 60-80 mesh Chromosorb W. As determined by means of a flame ionization detector, the method permits one to analyze phenol and cresols separately in urine at concentration levels ranging from 1 to 1000 mg/liter or 1 to 1000 ppm. With the column temperature and hydrogen carrier gas flow rate maintained at 150°C and 100 ml/min, respectively, the retention times for phenol and (m+p)cresol (unresolved) were approximately 3.1 and 3.9 min, respectively. For a urine sample containing about 30 ppm (w/w) of phenol, a standard deviation of 1.2 mg/liter was found (about 4% relative), whereas the sensitivity was sufficient to detect less than 1 ppm of phenol in 5 μl of urine. Furthermore, based on correlations between phenol excretion in urine and benzene exposure concentrations in the atmosphere, the GC results indicate that the limit of 200 mg of phenol/liter of urine could be considered medically safe.

Of greater significance are the techniques developed by Dalgliesh et al. [7] and Horning and Horning [8] for the separation of a wide range of biological metabolites occurring in urine and tissue extracts by gas chromatography and combined gas chromatography-mass spectrometry. These patterns are normally referred to as "human metabolic profiles."

Both reports are similar in content, scope, and instrumentation employed, since the studies were performed essentially by the same university group, the personnel of Baylor University's Institute of Lipid Research and Department of Biochemistry. Since urine contains a vast amount of substances representing intermediates and end products of metabolic pathways of both endogenous and exogenous compounds, the possibility of performing multicomponent analyses led directly to the concept of metabolic profile [8] analyses for the study of human metabolites. Dalgliesh et al. and Horning et al. developed GC procedures which permit a wide range of substances occurring in urine or tissue extracts to be separated and identified unequivocally by mass spectrometry. As noted by Horning and Horning, these analyses depend on several critical procedural steps:

1. The isolation of a suitable analytical sample from urine, blood, or other human tissue, sample isolation usually being accomplished by one or more procedures involving solvent extraction, ion exchange, column or thin-layer chromatography following in some instances special enzymatic or chemical hydrolysis steps [8].

2. The conversion of the isolated sample components into volatile, stable derivatives suitable for GC separation, including the formation of methyl esters, trimethylsilyl ethers, trimethylsilyl esters, acetylation, propionylation, methoxime formation, etc. As stressed in both studies, "derivative formation is usually a matter that must be studied in detail for each biological application. Many human metabolic products are polyfunctional and not readily volatilized; it is at times necessary to carry out two or three successive or simultaneous reactions in order to obtain suitable derivatives. Polyfunctional compounds may yield multiple products under some circumstances; this should be avoided whenever possible. Special derivatives may be needed for specific or selective detection" [8].
3. The GC analysis of the sample by a temperature-programmed separation with columns of relatively high efficiency.

Dalgliesh et al. used a Barber-Colman model 5000 gas chromatograph equipped with flame ionization detection and 6-ft x 3.5-mm-i.d. glass U-shaped columns packed with either 1% or 10% F-60 siloxane polymer coated on Gas Chrom P, 80-100 mesh, for the separation of the many metabolic products isolated from biological media, including:

1. Carboxyl groups that are quantitatively converted to methyl esters or trimethylsilyl esters while phenolic, alcoholic, and potential enolic groups are converted into trimethylsilyl esters. For qualitative studies, propionyl derivatives of hydroxy compounds may be formed. Regardless of the derivative employed, the column separations are achieved by temperature programming at 2°C/min over that portion of the 30 to 260°C temperature range that might be required.
2. The method is applicable to: (a) fatty acids above butyric; (b) di- and tricarboxylic acids; (c) hydroxy acids and keto acids; (d) polyhydroxy and alicyclic compounds such as glycerol, inositol, quinic acid, shikimic acid, ascorbic acid, and sugar alcohols; (e) aromatic hydroxy and acidic compounds, both benzenoid and indolic; (f) sesquiterpenes; (g) steroids; (h) glycine conjugates; (i) mercapturic acids; and (j) glucuronides. It is not satisfactory for sulfate conjugates, iminazoles, or polypeptides.

Although methylene units provide a useful, reproducible parameter for characterizing peak position, the nature of the derivatives was confirmed by integrated gas chromatography-mass spectrometry.

Horning and Horning used 12-ft packed columns with thermostable phases (SE-30, OV-1, OV-101, OV-17) and temperature programming (1 or 2°C/min) to develop metabolic profiles for human adult (Fig. 2.26) and infant (Fig. 2.27) urinary steroids, for serum (or plasma) and urinary sugars and sugar alcohols, for serum and urinary acids, for serum acids of the Krebs cycle (Fig. 2.28) and related compounds, and for many acidic and neutral drugs and metabolites.

In contrast to these high-molecular-weight metabolites requiring isolation and derivatization prior to GC analysis, Zlatkis and co-workers [219, 222,223] and Teranishi et al. [185] were primarily concerned with volatiles in heated urine specimens that could be isolated by purging with an inert gas stream.

Zlatkis and Liebich [219] studied by gas chromatography the volatile constituents of 30 urines from adults and also two by mass spectrometry. However, in this instance, rather than adsorbing the urinary headspace

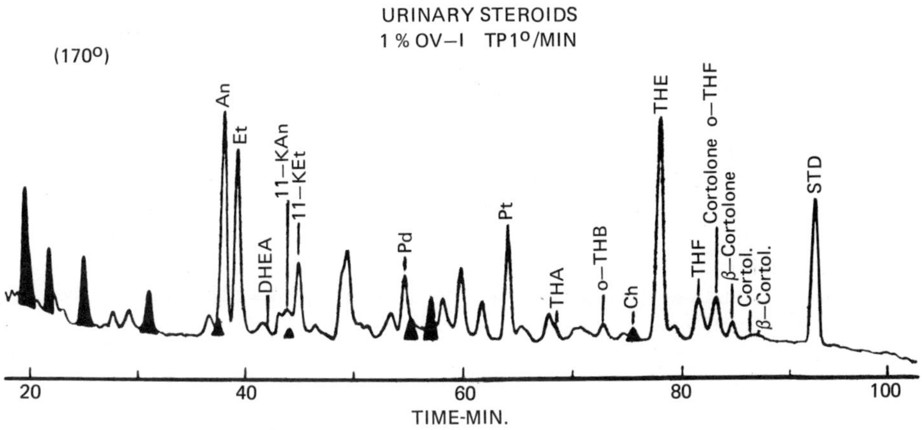

Figure 2.26. Urinary steroid profile (24-hr collection) for an adult male. The MO-TMSi and TMSi derivatives were separated with a 12-ft 1% OV-1 (on 100/120 mesh Gas Chrom P) column by temperature programming from 170°C at a rate of 1°C/min. The compounds noted in the figure are derivatives of androsterone (An), etiocholanolone (Et), dehydroepiandrosterone (DHEA), 11-ketoandrosterone (11-KAn), 11-ketoetiocholanolone (11-KEt), pregnanediol (Pd), pregnanetriol (Pt), tetrahydro-11-dehydrocorticosterone (THA), allo-tetrahydrocorticosterone (a-THB), cholesterol (Ch), tetrahydrocortisone (THE), tetrahydrocortisol (THF), cortolone, allo-tetrahydrocortisol (a-THF), β-cortolone, cortol, β-cortol, and cholesteryl butyrate (STD). The sample was obtained by enzymic (Glusulase) hydrolysis; the solid peaks represent "blank" substances arising from the enzyme preparation. From Horning and Horning [8], courtesy of Journal of Chromatographic Science.

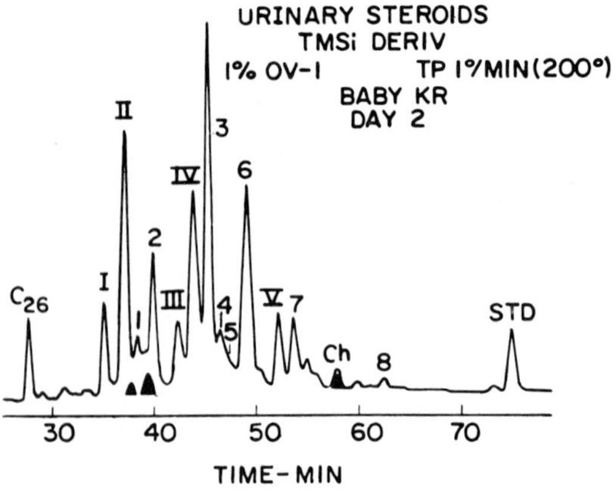

Figure 2.27. Urinary steroid profile (24-hr collection) for a newborn infant on the second day of life. The TMSi derivatives were separated in the same way as for Figure 2.26, starting at 200°C. The compounds noted in the figure are 5-androsten-3β,16α-diol-17-one (I), 5-androsten-3β,17β-diol-16-one (II), 5-androsten-3β,16α,17β-triol (III), 5-pregene-3β,16α-diol-20-one (IV), 5-pregene-3β,21-diol-20-one (V), compounds 1-8 are unidentified but include several triolones and tetrols in the 5-androsten-3β-ol series; the reference compounds are n-hexacosane (C_{26}) and cholesteryl butyrate (STD). The sample was obtained by enzymic (Glusulase) hydrolysis, and the solid peaks represent "blank" peaks derived from the enzyme preparation. From Horning and Horning [8], courtesy of Journal of Chromatographic Science.

volatiles on Tenax GC followed by desorption and GC analysis, the sample was prepared for GC-MS analysis by extracting 450 ml of sodium bicarbonate-neutralized urine sample for 24 hr with 80 ml of anhydrous diethyl ether in a continuous liquid-liquid extractor, distilling and trapping the volatiles from the extract under reduced pressure, and then concentrating the distillate to a final volume of nearly 15 μl. These concentrates of urinary volatiles from men and women were analyzed on a LKB 9000 gas chromatograph-mass spectrometer at 70 eV, with an ion-source temperature of 250°C and a jet-separator temperature of 210°C. The GC column used with the integrated GC-MS instrument was 210 m x 0.75-mm i.d. stainless-steel, coated with Dowfax 9N15.

On the other hand, 2.5-μl aliquots of all samples were introduced into a Perkin-Elmer model 900 gas chromatograph by direct injection, with use of a glass-lined insert, and separated on a 150-m x 0.5-mm-i.d. SS column coated with Dowfax 9N15. These high-efficiency capillary columns

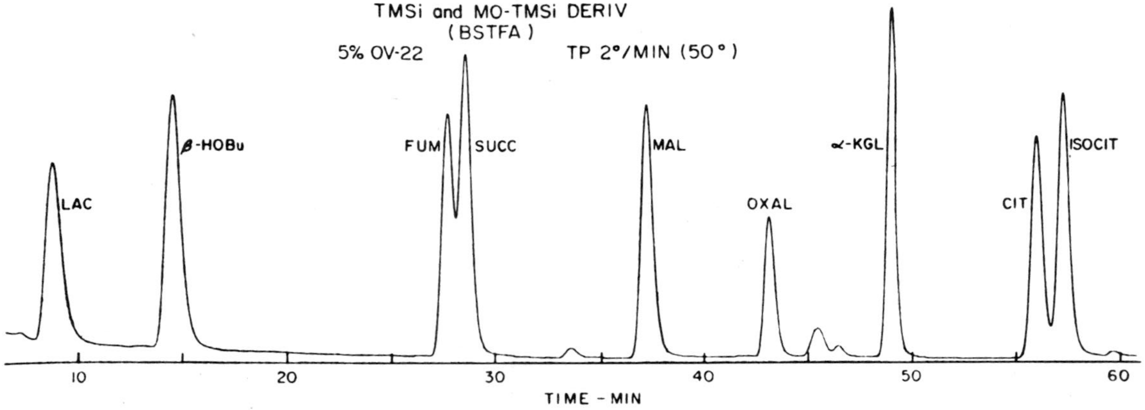

Figure 2.28. Separation of reference compounds as TMSi and MO-TMSi derivatives by temperature programming at 2°C/min from 50°C with a 12-ft 5% OV-22 column. The compounds are lactic acid (LAC), β-hydroxybutyric acid (β-HOBu), fumaric acid (FUM), succinic acid (SUCC), malic acid (MAL), oxalacetic acid (OXAL), α-ketoglutaric acid (α-KGL), citric acid (CIT), and isocitric acid (ISOCIT). From Horning and Horning [8], courtesy of Journal of Chromatographic Science.

provided the resolution necessary to indicate the presence of some 300 components when the column was programmed from ambient temperature to 150°C at 2°C/min. Of these 300 components, 40 were identified by mass spectrometry, including ketones, alcohols, lactones, terpenes, furans, dimethyl sulfone, pyrrole, and allyl isothiocyanate. Since dietary changes had only a relatively small effect on the profile of the volatiles and interindividual patterns were very similar, Zlatkis and Liebich theorized that these facts make analyses of this kind of potential diagnostic use.

In 1973, Zlatkis et al. [222] undertook the study of volatile metabolites in human urine with the object of examining qualitative and quantitative variations in the metabolic patterns. Using the procedure developed by Zlatkis et al. [186] (see breath analysis section for details) for sampling, transfer, and analysis of urine volatiles by gas chromatography, individual profiles were observed over a period of 2 months. As in previous studies involving dietary changes, no characteristic variations in the chromatograms could be observed. Furthermore, similar, highly constant, and reproducible results were also found in the profiles of sulfur components detected with a specific flame photometric sulfur detector. Characteristic constituents in normal urines included such compounds as 2-butanone, 2-pentanone, 4-heptanone, dimethyl sulfide, several alkyl furans, pyrrole, and carvone. Whereas pyrazines are present in trace quantities, dimethyl sulfone, a major constituent identified in urine extracts, was not found by the adsorption procedure. On the other hand, in subjects with diabetes mellitus under insulin treatment, high concentrations of pyrazines, cyclohexanone, lower aliphatic alcohols, and octanols were found. They noted that further studies will be required to investigate the correlations between the volatile metabolites produced in the different forms and stages of diabetes mellitus and in the different forms of treatment. Simpler chromatographic profiles can be obtained by selective detectors (flame photometric, electron capture) and subtractive techniques.

In a subsequent study of volatile urinary metabolites where three types of profiles were required (flame, sulfur, and GC-MS identifications), Zlatkis et al. [223] identified 51 compounds by integrated GC-MS (Table 2.20). A comparison of volatile urinary organic sulfur compounds from normal and diabetic individuals monitored by the flame photometric sulfur detector did indeed yield simpler but widely different chromatographic profiles as shown in Figure 2.29. In a single-blind study, they found that different forms of diabetics have distinctly different profiles, and it is anticipated that volatile organic analyses will add additional information to the comparative physiology of diabetic individuals.

Witten and co-workers [227] also studied by integrated gas chromatography-mass spectrometry the urinary acid metabolic profiles of 26 healthy young adults who were maintained on a palatable standard diet for 3 days. Specifically, they studied the direction and magnitude of changes in those excretion rates following an oral dose of ethanol. Following the

TABLE 2.20

Volatile Organic Compounds in Normal Human Urine[a]

Compound	Compound
Acetone	C_4-Furan
2-Butanone	Thiolan-2-one (tent.)
Ethanol	2-Heptanone
Propionaldehyde	3-Methylcyclopentanone
3-Methyl-2-butanone	6(?)-Methyl-3-heptanone
2,3-Butanedione	3-Octanone
2,3-Dimethylfuran	Methylpyrazine
2-Pentanone	2-n-Pentylfuran
2,4-Dimethylfuran	2,6- or 2,5-Dimethylpyrazine
Chloroform	Allylisothiocyanate
4-Methyl-2-pentanone	2,3-Dimethylpyrazine
3-Methyl-2-pentanone	Furfural
2(?)-Methyl-5(?)-ethylfuran	2,3,5-Trimethylpyrazine
Toluene	Vinylpyrazine
Dimethyldisulfide	2-Methyl-6-ethylpyrazine
3-Hexanone	Pyrrole
2,3,5-Trimethylfuran	2-Nonanone
5-Methyl-3-hexanone	Acetylfuran
3-Penten-2-one	p-Methylpropenylbenzene (tent.)
1-Methylpyrrole	Benzaldehyde
4-Methyl-3-penten-2-one (tent.)	β-Pinene (tent.)
1-Butanol	2-Methyl-6(?)-vinylpyrazine
4-Heptanone	2-Methylpyrrole
	Dimethylpyrrole
	1-Butylpyrrole (tent.)
	Carvone
	Piperitone
	p-Cresol

[a]From Zlatkis et al. [223], courtesy of Journal of Chromatographic Science.

ingestion of ethanol in a dosage of 1.2 g/kg of body weight over a 30-min period, the urines were extracted by a procedure which was a modified version of the method proposed by Hammond and Goodman [232]. Trimethylsilylation, quantitative gas chromatography, and mass spectrometry

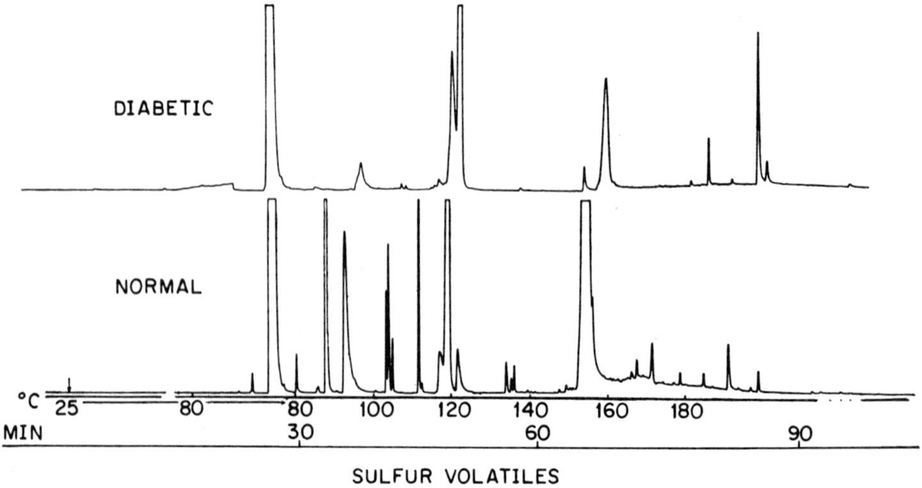

Figure 2.29. Comparison of urinary volatile organic sulfur compounds from normal and diabetic individuals. From Zlatkis et al. [223], courtesy of Journal of Chromatographic Science.

were performed as previously described [233,234]. After the ingestion of ethanol, means and standard deviations of the excretion rates of individual, identified acids were determined and compared to basal values as shown in Table 2.21, where it is noted that there is a significant net effect of ethanol for only six of the acids studied: α- and β-hydroxybutyric, adipic, β-methyladipic, p-hydroxyphenylacetic, and 2,5-furandicarboxylic acids. In addition to these findings, Witten et al. point out that the metabolic effects of ethanol have been studied in a number of systems. For example, there is evidence that the tricarboxylic acid cycle [235], hepatic gluconeogenesis [236], and oxidative deamination [237] are inhibited by ethanol, whereas acetaldehyde, a metabolite of ethanol, inhibits catecholamine degradation [238]. Also, other studies have shown that hormonal release from the adrenal medulla and cortex [239] is more stimulated in "naive" drinkers than in chronic alcoholics [240].

Teranishi and co-workers [185] developed a technique of direct GC analysis of volatiles from human breath and urine, and this concept was extended by Robinson et al. [170], who designed a completely automated apparatus for the quantitative analysis of volatile compounds in urine vapor suitable for the routine determination of about 200 volatile constituents of human urine. In their arrangement, a 50-ml sample of urine, 19.65 g of KH_2PO_4, and 30.35 g of Na_2HPO_4 are placed in each of four sample bottles and a key is depressed on the teletype of a PDP 11/20 computer. The entire analysis is then carried out by the apparatus without any manual operations in approximately 6 hr.

Using this apparatus to develop diagnostic procedures for various human diseases, Matsumoto et al. [228] to date have identified by GC-MS 52 compounds, using a modified headspace procedure whereby the organic components in urine are separated from the water-saturated sweep gas by adsorbtion onto a Chromosorb 101 precolumn [241]. The volatiles are transferred to a stainless steel helical trap cooled with liquid nitrogen by backflushing the Chromosorb trap for 90 min. The sample is flashed from the helical trap onto the chromatographic column with a heat gun. Preliminary compound identification was made using Cornu and Massot's compilation of mass spectra [242,243], but final identities are confirmed, in most cases, by comparisons of both relative retention times and mass spectra with those of authentic compounds obtained on the same GC-MS system.

Using concentration techniques similar to those reported by Matsumoto et al. [228] and Zlatkis and co-workers [186] where trace volatile constituents of human urine, serum, and cerebrospinal fluid are adsorbed on a porous polymer precolumn, Novotny et al. [6] resolved these volatiles with high-efficiency glass capillary columns coated with SF-96 methylsilicone oil or Emulphor ON-870 which yielded film thicknesses between 0.1 and 0.2 μm.

To study the volatile by-products resulting from cleavage of urinary sulfates and glucuronides, they modified the method of Horning and Horning [8] as applied to urinary steroids. A 100-ml urine sample, to which a 0.2-ml aliquot of Glusulase (a mixture of sulphatase and β-glucuronidase) has been added, is incubated at 37°C for 18 hr, with gentle stirring. To this incubated sample, potassium hydroxide (400 g/liter) is added to adjust the sample's pH to 5.5, followed by another 0.2 ml of enzyme and an 18-hr incubation period. A 50-ml portion of this urine is then processed for analysis by the same procedure as that used for untreated urine.

Using a gas chromatograph equipped with a flame ionization detector and a modified injection system similar to that described by Zlatkis et al. [186], all chromatograms are obtained using either a 55-m or 80-m x 0.31-mm-i.d. glass capillary column coated with either SF-96 or Emulphor ON-870; the injector and detector temperatures used are 240 and 265°C, respectively. As noted by Novotny et al., columns of different polarities prove useful for the resolution of many isomeric compounds. This is indicated by the data in Table 2.22, which compares the elution temperatures for reference compounds on these two stationary phases.

C. Blood and Tissue Analysis [7,8,66-69,74,80,82,83,85-87, 90,91,98,100,101,104,106-108,110,111,114,118,119,130,131,133-135, 139,142,143,146,147,150,184,211,212,244-251]

We have seen that many of the methods developed for ethanol analysis have been used as is or with modifications for the determination of other volatiles

TABLE 2.21

Mean Excretion Rate Difference of Urinary Acids for All Subjects, Standard Deviation of the Differences, and Level of Significance[a,b]

Identity	Excretion rate difference paired by subject (μg/mg of creatinine)					
	Activity - basal			(Activity + ethanol) - basal		
	Mean	Std. dev.	α(A - B)[c]	Mean	Std. dev.	α[(A + E) - B][d]
α-Hydroxybutyric	Not detectable	-	-	6.2	12	0.05
β-Hydroxybutyric	4.5	27	N.S.[e]	21	21	0.001
β-Hydroxyisovaleric	2.5	3.3	N.S.	1.4	10.8	N.S.
Internal standard-methylmalonic	-	-	-	-	-	-
Succinic	16	12	0.05	16	21	0.01
Unidentified, 1	-0.097	0.90	N.S.	-0.031	3.3	N.S.
Unidentified, 2	9.8	24	N.S.	4.0	4.3	0.001

Adipic	-0.38	1.6	N.S.	4.1	6.7	0.01
β-Methyladipic	2.9	4.3	N.S.	13	11	0.001
5-Hydroxymethyl-2-furoic	39	42	0.05	20	21	0.001
Tartaric	9.0	7.6	0.05	12	8.7	0.001
p-Hydroxyphenylacetic	-2.1	3.1	N.S.	8.7	6.9	0.001
2,5-Furandicarboxylic	0.75	7.1	N.S.	12	9.8	0.001
"Envelope" containing aconitic (O), citric & isocitric (P), and tetronic (N - P)	-	-	-	-	-	-
Dihydroxyphenylproprionic	-1.8	15	N.S.	-1.7	17	N.S.

[a]From Witten er al. [227], courtesy of Clinical Chemistry.
[b]Calculated from the Student's t-distribution. Indicates a significant net effect of ethanol.
[c]Level of significance of the mean difference owing to activity.
[d]Level of significance of the mean difference owing to activity + ethanol.
[e]Difference is not significant ($\alpha > 0.5$).

TABLE 2.22

Comparison of Elution Temperatures for Reference Compounds on Two Stationary Phases[a]

Compound	Elution temp. (°C)	
	On SF-96	On Emulphor
Cyclohexanone	59	79
Benzaldehyde[b]	61	93
Acetophenone	73.5	104
2-Octanone[c]	82	81
Undecane	90	79
Carvone[d]	100	-
Tridecane	113	95
1-Octanol	-	106
Pentadecane	137	119
1-Decanol	-	130
Heptadecane	148	131
Octadecane	170	151
1-Dodecanol	-	152
Eicosane	190	175

[a]From Novotny et al. [6], courtesy of Clinical Chemistry.

The following columns and conditions were used for these measurements:

(1) 55-m x 0.31-mm-i.d. Pyrex capillary coated with SF-96, programmed from 35 to 80°C at 2°C/min, and then from 80 to 210°C at 1°C/min.

(2) 38-m x 0.23-mm-i.d. soft-glass capillary coated with Emulphor ON-870, programmed from 70 to 190°C at 1°C/min.

[b]Approximate elution temperature for the same compound 150°C in ref. 223 and 110 to 115°C in ref. 228.

[c]Approximate elution temperature for 3-octanone between 100 and 140°C in ref. 223 and 110 and 115°C in ref. 228.

[d]Approximate elution temperature for the same compound 180°C in ref. 223 and 160°C in ref. 228.

present in biological media. Instruments described for analyzing breath samples and urine specimens are also applicable to blood, cerebrospinal fluid, and other body tissues. Those not previously mentioned in the text will be discussed briefly in this section.

In 1961, Hill and Butler [244] used gas chromatography for the detection of toxic gases in tissues by direct sampling of the air taken in a sealed container. Soon thereafter, Curry et al. [130] reported their findings using an apparatus such as that shown in Figure 2.30 which proved suitable for the detection of common volatile compounds such as ether, halothane, chloroform, carbon tetrachloride, ethanol, acetone, trichloroethylene, and paraldehyde. Essentially a headspace gas sampler, a 1-ml blood sample is injected through a silicone rubber septum into a closed glass cylinder which contains 0.5 g of solid K_2CO_3. The container (about 7 ml in volume) is shaken and then immersed for 1 min into a 70°C water bath (the K_2CO_3 serving to reduce the water vapor pressure while at the same time by its heat of solution providing its own internal heating). A 1-ml sample of the headspace vapors is withdrawn and injected into the gas chromatograph equipped with flame ionization and a 12-ft column packed with 10% silicone oil on 60-80 mesh Embacel; the column being maintained at room temperature. If paraldehyde is suspected, it can be converted to acetaldehyde by substituting 1 ml of sulfuric acid for the K_2CO_3. Under these conditions, acetaldehyde has a retention time that is almost equivalent to that of ether.

Lester [245] determined acetic acid in blood and other tissues by first separating the acid from a copper sulfate-saturated perchloric acid filtrate of blood or other tissue by incomplete vacuum distillation in which the course of the distillation could be represented by the relationship [252]

$$\log \frac{y_i}{y_f} = 15.7 \log \frac{x_i}{x_f} \tag{2.26}$$

where

y_i, y_f = the initial and final volumes of filtrate, respectively
x_i, x_f = the initial and final amounts of acetic acid in the filtrate, respectively

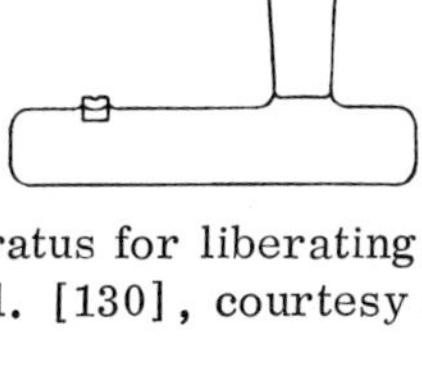

Figure 2.30. A glass apparatus for liberating volatile organics from blood samples. From Curry et al. [130], courtesy of Nature.

The vacuum distillation apparatus with its 15-ml distilling flask and 5-ml receiver are shown in Figure 2.31. Five microliters of the distillate are injected onto a 1.5-m x 3-mm-o.d. glass column packed with 8% Carbowax 20M and 2% terephthalic acid on 20-80 mesh Fluoropak. The column is maintained at 130°C and the injector at 180°C. The carrier gas (nitrogen) is passed through a 30-cm length of 9.53-mm-o.d. copper tubing packed with 60-80 mesh Chromosorb and saturated with formic acid; this tube is kept at ambient temperature. Capable of detecting an acetic acid concentration of 0.1 mg of acetic acid/liter of filtrate, Lester found that the normal values of free acetate per liter of human plasma were 2 to 12 mg, increasing to 75 mg/liter after alcohol ingestion; similar elevated concentrations were found after ingestion of about 8 g of sodium acetate. No peaks were noted by Lester from pyruvate, lactate, β-hydroxybutyrate, or acetoacetate when carried through the complete procedure.

Modifying a previous method incorporating a solvent-extraction procedure for the determination of blood ether levels by GC, Northington and Owens [246] described a procedure which was applicable for the determin-

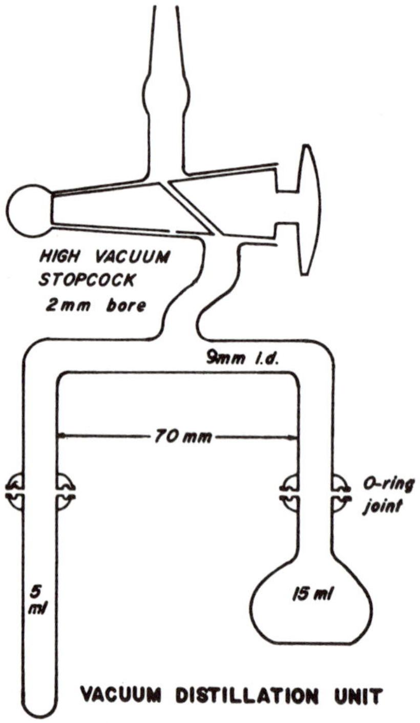

Figure 2.31. Vacuum distillation apparatus. From Lester [245], courtesy of Analytical Chemistry.

ation of ether levels in brain tissue. With a Beckman GC-2 gas chromatograph equipped with a thermal conductivity detector, ether recovery studies showed weight differences between ether added to brain tissue and ether recovered in cyclohexane ranging from 0.00 to 0.23 mg with an average difference of ±0.06 mg. Over 50% showed a difference of 0.05 mg or less, while second extractions performed on randomly selected samples showed that the initial recovery was within the range 96 to 98%. The results for 24 brain samples taken from 21 mice were consistent with those reported by Dybing and Dybing [253] for studies done on the ether content of the brains of rabbits following ether-induced spontaneous respiratory paralysis.

Baker and co-workers [247] determined by GC the simultaneous analysis of lower alcohols, acetone, and acetaldehyde in blood by direct injection of whole blood into the chromatograph without prior processing. Using a Micro-Tek 220 equipped with a flame ionization detector, the separation was performed by a 183-cm x 5-mm-i.d. glass column filled with 60-80 mesh Porapak Q whose temperature was held at 110°C; the temperatures of the inlet and detector were set at 125°C. With a nitrogen carrier gas flow rate of 35 ml/min, the relative retention times of blood volatiles compared to acetonitrile were water, 0.12; methanol, 0.29; acetaldehyde, 0.44; ethanol, 0.73; acetonitrile, 1.00; acetone, 1.50; isopropanol, 1.70; and n-propanol, 2.40. As noted by Baker et al., the use of glass inlets (interchangeable inserts) minimized decomposition of the sample and facilitated cleaning of the inlets. The inlets were easily removed, cleaned, and replaced after 25 μl of whole blood had been injected.

The direct-injection method proposed by Jain [110] for the determination of blood alcohols and other volatile compounds (for details, see ethanol analysis by direct injection) is also capable of determining compounds such as those listed in Table 2.23, in which their retention times relative to ethanol are compared at two different column temperatures.

Recently, Stoner, Cowburn, and Craig [251] devised a technique for the isolation and concentration of neutral organic volatiles prior to their examination by gas chromatography.

With 150 μg of naphthalene added as internal standard to 15 ml of plasma, three 20-ml portions of ether are used to extract the organics from the plasma. Following each ether addition and shaking, the upper ether phase is removed and pooled. The extracts are concentrated on a special fractionation column (Fig. 2.32), the concentration proceeding until the ether volume is reduced to about 0.8 ml. The concentrate is then transferred to 6-mm x 50-mm glass tubes and further evaporated in a slow stream of dry nitrogen at 4°C to a final volume approaching 100 μl. Ten microliters of the concentrate are then injected into an F&M model 402 gas chromatograph equipped with flame ionization detection and 2.5-m x 2-mm-i.d. glass columns packed with 10% Carbowax 1500 coated on 100-120 mesh inert solid support. With a helium carrier gas flow rate of

TABLE 2.23

Retention Data for Some Compounds[a,b]

	Oven temp. 100°C		Oven temp. 130°C	
	Retention time (min)	Retention time relative to ethanol	Retention time (min)	Retention time relative to ethanol
Acetaldehyde	2.56	0.73	1.25	0.60
Acetone	2.38	0.68	1.62	0.76
Benzene	4.31	1.23	2.81	1.33
Chloroform	5.62	1.61	3.25	1.53
Ethanol	3.50	1.00	2.12	1.00
Formaldehyde	1.81	0.52	1.38	0.65
Isobutanol	7.25	2.10	3.94	1.86
Isopropanol	3.19	0.91	2.06	0.97
Methanol	3.25	0.93	2.06	0.97
Methyl ethyl ketone	3.38	0.97	2.19	1.03
Propyl acetate	4.50	1.29	2.75	1.30
Toluene	6.94	1.98	4.06	1.92
Xylene	11.38	3.25	6.38	3.01

[a]From Jain [110], courtesy of Clinical Chemistry.

[b]6-ft x 0.25-in. SS column packed with 30% Carbowax 20M on Chromosorb W, AW, 60/80 mesh. Flow rates: nitrogen, 35 ml/min; hydrogen, 28 ml/min; oxygen, 100 ml/min.

35 ml/min, the other temperature conditions employed were flash heater, 85°C; detector, 95°C; column, programmed as follows: postinjection isothermal at 55°C for 20 min, then programmed linearly at 10°C/min to 115°C, this latter temperature of 115°C then being maintained for 60 min.

With this method, the recovery of naphthalene in the initial extract was in excess of 95%. The concentration of components detected by Stoner et al. was in the range 1 to 100 nmoles. Peak positions and relative detection sensitivities of some standards are shown in Figure 2.33. Stoner et al.

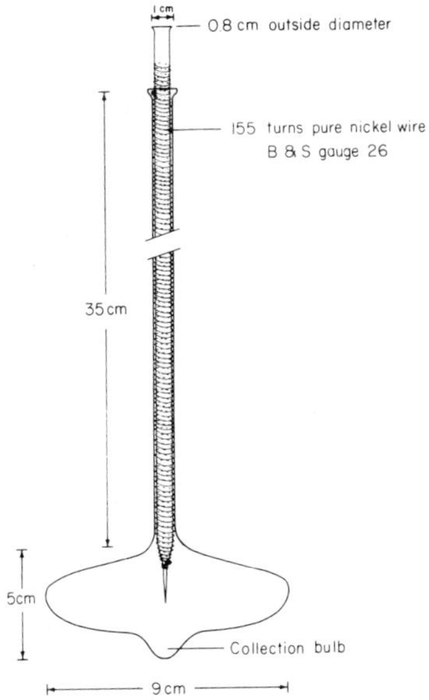

Figure 2.32. Apparatus for concentration of volatiles from ether. From Stoner et al. [251], courtesy of Analytical Chemistry.

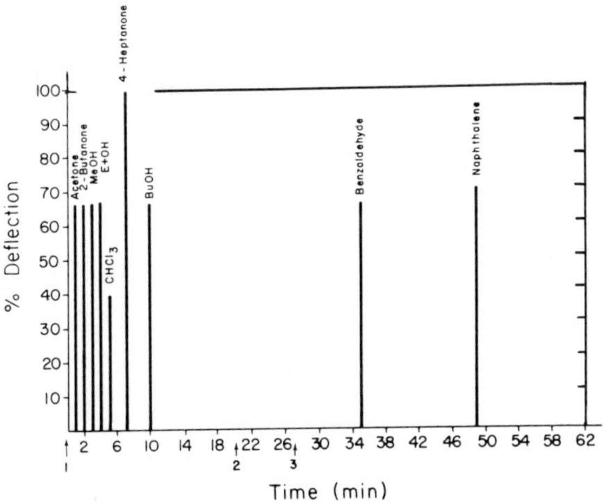

Figure 2.33. Sensitivity of standards in GC procedure. Range 100, attenuation 1 on instrument described in text. Ambient temperature (55°C), program: 10°/min to 115°C limit, isothermal at 115°C. 100 nmoles of each compound applied. From Stoner et al. [251], courtesy of Analytical Chemistry.

also point out that actual structural determination of the 23 or more volatile components partially resolved by this method, in normal and disease states, was currently under way, using mass spectrometry and other methods.

REFERENCES

1. Goth, A., Medical Pharmacology, 7th ed., C. V. Mosby, St. Louis, 1974.
2. Rubin, E. and Lieber, C. S., Science, 172, 1097 (1971).
3. Forney, R. B. and Hughes, F. W., Combined Effects of Alcohol on Other Drugs, Thomas, Springfield, Ill., 1968.
4. Weiner, M., New Engl. J. Med., 283, 871 (1970).
5. Sundby, P., Alcoholism and Mortality, Universitetsforlaget, Oslo, 1967.
6. Novotny, M., McConnell, M. L., Lee, M. L., and Farlow, R., Clin. Chem., 20, 1105 (1974).
7. Dagleish, C. E., Horning, E. C., Horning, M. G., Knox, K. L., and Yarger, K., Biochem. J., 101, 792 (1966).
8. Horning, E. C. and Horning, M. G., J. Chromatogr. Sci., 9, 129 (1971).
9. Ritchie, J. M., in L. S. Goodman and A. Gilman (Eds.) The Pharmacological Basis of Therapeutics, 4th ed., Macmillan, New York, 1970.
10. Westerfeld, W. W., Amer. J. Clin. Nutr., 9, 426 (1961).
11. Conney, A. H., Pharmacol. Rev., 19, 317 (1967).
12. Schenkman, J. B., Biochemistry, 9, 2081 (1970).
13. Svoboda, D. J. and Manning, R. T., Amer. J. Pathol., 44, 645 (1964).
14. Ariyoshi, T., Takabatake, E., and Remmer, H., Life Sci., 9, 361 (1970).
15. Wada, O., Yuzo, Y., Gumoei, U., and Kiku, N., Biochem. Pharmacol., 17, 595 (1968).
16. Shanley, B. C., Zail, S. S., and Joubert, S. M., Lancet, 1, 70 (1968).
17. Zieve, L., Hill, E., Schwartz, S., and Wilson, C. J., J. Lab. Clin. Med., 41, 663 (1953).
18. De Matteis, F., Pharmacol. Rev., 19, 523 (1967).
19. Marver, H. S., Collins, A., Tscguddy, D. P., and Reichigl, M., Jr., J. Biol. Chem., 241, 4323 (1966).
20. Shanley, B. C., Zail, S. S., and Joubert, S. M., Brit. J. Haematol., 17, 389 (1969).
21. Mendelson, J. H., Stein, S., and Mello, N. K., Metabolism, 14, 1255 (1965).

22. Kater, R. M. H., Roggin, G., Tobon, F., Zieve, P., and Iber, F. L., Amer. J. Med. Sci., 258, 35 (1969).
23. Orme-Johnson, W. H. and Ziegler, D. M., Biochem. Biophys. Res. Commun., 21, 78 (1965).
24. Isselbacher, K. J. and Carter, E. A., Biochem. Biophys. Res. Commun., 39, 530 (1970).
25. Keillin, D. and Hartree, E. F., Biochem. J., 39, 293 (1945).
26. Roach, M. K., Reese, W. N., and Creaven, P. J., Biochem. Biophys. Res. Commun., 36, 596 (1969).
27. Tephly, T. R., Tinelli, F., and Watkins, W. D., Science, 166, 627 (1969).
28. Papenberg, J., von Wartburg, J. P., and Aebi, H., Enzymol. Biol. Clin., 11, 237 (1970).
29. Gigon, P. L., Gram, T. E., and Gillette, J. R., Mol. Pharmacol., 5, 109 (1969).
30. Sasame, H. A. and Gillette, J. R., Biochem. Pharmacol., 19, 1025 (1970).
31. Sasame, H. A. and Gillette, J. R., Mol. Pharmacol., 5, 123 (1969).
32. Jones, A. L. and Armstrong, D. T., Proc. Soc. Exp. Biol. Med., 119, 1136 (1965).
33. Iseri, O. A., Lieber, C. S., and Gottlieb, L. S., Amer. J. Pathol., 48, 535 (1966).
34. Rubin, E., Hutterer, F., and Lieber, C. S., Science, 159, 1469 (1968).
35. Rubin, E., Bacchin, P., Gang, H., and Lieber, C. S., Lab. Invest., 22, 569 (1970).
36. Rubin, E. and Lieber, C. S., Gastroenterology, 52, 1 (1967).
37. Rubin, E. and Lieber, C. S., New Engl. J. Med., 278, 869 (1968).
38. Lieber, C. S. and Rubin, E., Amer. J. Med., 44, 200 (1968).
39. Rubin, E., Lieber, C. S., Alvares, A. P., Levin, W., and Kuntzman, R., Biochem. Pharmacol., 20, 229 (1971).
40. Rubin, E. and Lieber, C. S., Science, 162, 690 (1968).
41. Misra, P. S., Lefevre, A., Ishii, H., Rubin, E., and Lieber, C. S., Amer. J. Med., in press, 1971.
42. Rubin, E., Misra, P. S., Gang, H., and Lieber, C. S., Amer. J. Med., 49, 801 (1970).
43. Lieber, C. S. and De Carli, L. M., Science, 162, 917 (1968).
44. Lieber, C. S. and De Carli, L. M., J. Biol. Chem., 245, 2505 (1970).
45. Lieber, C. S., Rubin, E., and De Carli, L. M., Biochem. Biophys. Res. Commun., 40, 858 (1970).
46. Lieber, C. S., Rubin, E., De Carli, L. M., Misra, P., and Gang, H., Lab. Invest., 22, 615 (1970).
47. Lieber, C. S. and De Carli, L. M., Science, 170, 78 (1970).

48. Rubin, E., Gang, H., and Lieber, C. S., Biochem. Biophys. Res. Commun., 42, 1 (1971).
49. Lieber, C. S. and De Carli, L. M., Clin. Res., 12, 274 (1964).
50. Davis, V. E. and Walsh, M. J., Science, 167, 1005 (1970).
51. Holtz, P., Stock, K., and Westermann, E., Nature, 203, 656 (1964).
52. Davis, V. E., Walsh, M. J., and Yamanaka, Y., J. Pharmacol. Exp. Ther., 174, 401 (1970).
53. Goldstein, A. and Judson, B. A., Science, 171, 290 (1971).
54. Seixas, F., in Chemical & Engineering News, p. 16, May 1, 1972.
55. Ahtee, L., in Chemical & Engineering News, pp. 14-15, May 1, 1972.
56. Tewari, S. and Noble, E. P., in Chemical & Engineering News, p. 14, May 1, 1972.
57. Calder, G. V., J. Chem. Educ., 51, 19 (1974).
58. Harger, R. N., in C. P. Stewart and A. Stolman (Eds.), Toxicology, Mechanisms and Analytical Methods, Academic Press, New York, 1961.
59. Cadman, W. J. and Johns, T., paper presented at the Ninth Annual Conference on Analytical Chemistry and Applied Spectroscopy, Pittsburgh, March 1958.
60. Cravey, R. H., Reed, D., and Sedgwick, P. R., Amer. Lab., pp. 63-71, May 1972.
61. Jain, N. C. and Cravey, R. H., J. Chromatogr. Sci., 10, 263 (1972).
62. Cravey, R. H. and Jain, N. C., J. Chromatogr. Sci., 12, 209 (1974).
63. Jain, N. C. and Cravey, R. H., J. Chromatogr. Sci., 12, 214 (1974).
64. Patterson, D. A., in H. Purnell (Ed.), New Developments in Gas Chromatography, Vol. 2, Wiley, New York, 1973.
65. Dubowski, K. M., Clin. Chem., 20, 294 (1974).
66. Cadman, W. J. and Johns, T., J. Forensic Sci., 5, 369 (1960).
67. Lyons, H. and Bard, J., Clin. Chem., 10, 429 (1964).
68. Davis, R. A., J. Forensic Sci., 11, 205 (1966).
69. Steinberg, M., Nash, J. B., and Walker, J. Q., J. Forensic Sci., 10, 201 (1965).
70. Osterhaus, F. and Johannsmeir, K., Blutalkohol, 2, 1 (1963).
71. Skrabvanek, P. and Novak, J., Blutalkohol, 3, 271 (1966).
72. Griffiths, W. C., Oleksyk, S. K., and Diamond, I., Clin. Biochem., 6, 124 (1973).
73. Rockerbie, R. A., Can. Pharm. J., 96, 38 (1963).
74. Laessig, R. H., Anal. Chem., 40, 2205 (1968).
75. Harger, R. N. and Forney, R. B., in A. Stolman (Ed.), Progress in Chemical Toxicology, Vol. 1, Academic Press, New York, 1963.
76. Gerarde, H. W., in Proceedings of International Symposium on Microchemical Techniques--August 16, 1961, Wiley, New York, 1962, p. 1009.
77. Gerarde, H. W. and Walker, A. R., Amer. J. Med. Tech., 28, 327 (1962).

78. Goldberg, L. and Rydberg, W., Automation in Analytical Chemistry, Mediad, New York, 1966.
79. Master, R., Barrais, C., and Legendre, G., Trav. Soc. Pharm. Montellier, 24, 68 (1964).
80. Curry, A. S., Walker, G. W., and Simpson, G. S., Analyst (London), 91, 742 (1966).
81. Bonnichsen, R. and Lundgren, G., Acta Pharmacol. Toxicol., 10, 223 (1954).
82. Roach, M. K. and Creaven, P. J., Clin. Chim. Acta, 21, 275 (1968).
83. Truitt, E. B. and Duritz, G., in R. P. Maickel (Ed.), Biochemical Factors in Alcoholism, Pergamon Press, Oxford, 1967.
84. Barrett, M. J., P-E Clin. Chem. Newsl., 3, 1 (1971).
85. Fox, J. E., Proc. Soc. Exp. Biol. Med., 97, 236 (1958).
86. McCord, W. M. and Gadsden, R. H., J. Gas Chromatogr., 2, 38 (1964).
87. Gadsden, R. H. and McCord, W. M., J. Gas Chromatogr., 2, 7 (1964).
88. Wesselman, H. J., J. Amer. Pharm. Ass., 49, 320 (1960).
89. Burchfield, H. P. and Storrs, E. E., Biochemical Applications of Gas Chromatography, Academic Press, New York, 1962.
90. Maricq, L. and Molle, L., Bull. Acad. Roy. Med., 24, 199 (1959).
91. Maricq, L. and Molle, L., Annales Pharm. Fr., 18, 811 (1960).
92. Aerograph Research Notes, Summer Edition, 1960, Wilkins Instrument and Research, Inc., Walnut Creek, Calif.
93. Chundela, B. and Janak, J., J. Forensic Med., 7, 153 (1960).
94. Chundela, B. and Janak, J., Cas. Lek. Ces., 99, 90 (1960).
95. Drawert, F., Felgenhauer, R., and Kupfer, G., Angew. Chem., 72, 385 (1960).
96. Drawert, F. and Kupfer, G., Hoppe-Seyler's Z. Physiol. Chem., 329, 90 (1962).
97. Machata, G., Mikrochim. Acta, 4, 691 (1962).
98. Parker, K. D., Fontan, C. R., Yee, L. L., and Kirk, P. L., Anal. Chem., 34, 1234 (1962).
99. Bonnichsen, R. and Linturi, M., Acta Chem. Scand., 16, 1289 (1962).
100. Coldwell, B. B. and Grant, G. L., Chem. in Canada, 15, 16 (1963).
101. Hessel, D. W. and Modglin, F. R., J. Forensic Sci., 9, 255 (1964).
102. Brahm-Vogelsanger, A. and Wagner, H. J., Dtsch. Z. Ger. Med., 55, 137 (1964).
103. Freudiger, J. B. and Viguau, J. A., J. Forensic Sci., 10, 73 (1965).
104. Mather, A. and Assimos, A., Clin. Chem., 11, 1023 (1965).
105. Sturner, W. Q. and Coumbis, R. J., Amer. J. Clin. Pathol., 46, 349 (1966).
106. Curry, A. S., Walker, G. W., and Simpson, G. S., Analyst, 91, 1088 (1966).

107. Freund, G., Anal. Chem., 39, 545 (1967).
108. Colehour, J. K., Anal. Chem., 39, 1190 (1967).
109. Cooper, J. D. H., Clin. Chim. Acta, 33, 483 (1971).
110. Jain, N. C., Clin. Chem., 17, 82 (1971).
111. Finkle, B. S., in I. Sunshine (Ed.), Manual of Analytical Toxicology, CRC Press, Cleveland, 1971.
112. Solon, J., Watkins, J., and Mikkelsen, L., J. Forensic Sci., 17, 447 (1972).
113. Bonnichsen, R., Hedjfall, B., and Ryhage, R., Z. Rechtsmedizin, 71, 134 (1972).
114. Sine, H. E., McKenna, M. J., Law, M. R., and Murray, M. H., J. Chromatogr. Sci., 10, 297 (1972).
115. Blume, P., Berchild, K. M., and Cawley, L. P., Anal. Biochem., 54, 429 (1973).
116. Baird, W. W., Pacific Conference on Chemistry and Spectroscopy, San Diego, November 1973.
117. Hartigan, M. J., Hoberecht, H. D., and Purcell, J. E., Amer. Lab., pp. 36-44, October 1973.
118. Brown, G. A., Neylan, D., Reynolds, W. J., and Smalldon, K. W., Anal. Chim. Acta, 66, 271 (1973).
119. Smalldon, K. W. and Brown, G. A., Anal. Chim. Acta, 66, 285 (1973).
120. Pereira, W. E., Summons, R. E., Rindfleisch, T. C., and Duffield, A. M., Clin. Chim. Acta, 51, 109 (1974).
121. Tenney, H. M. and Harris, R. J., Anal. Chem., 29, 317 (1957).
122. Drawert, F. and Kupfer, G., Angew. Chem., 72, 33 (1960).
123. Smith, H. W., J. Lab. Clin. Med., 38, 762 (1951).
124. McNair, H. M. and Bonelli, E. J., Basic Gas Chromatography, Consolidated Printers, Berkeley, Calif., 1968.
125. Ladd, S. L. and Wallace, J. E., in W. W. Stewart (Ed.), Drug Abuse in Industry, Halos and Associates, Miami, Fla., 1970.
126. Winek, C. L., Clin. Toxicol., 3, 541 (1970).
127. Sunshine, I. (Ed.), Handbook of Analytical Toxicology, Chemical Rubber Co., Cleveland, 1969.
128. Clarke, E. G. C., Isolation and Identification of Drugs, Pharmaceutical Press, London, 1969.
129. Reynolds, W. E., Bacon, V. A., Bridges, J. C., Coburn, T. C., Halpern, B., Lederberg, J., Levinthal, E. C., Steed, E., and Tucker, R. B., Anal. Chem., 42, 1122 (1970).
130. Curry, A. S., Hurst, G., Kent, N. R., and Powell, H., Nature, 195, 603 (1962).
131. Duritz, G. and Truitt, E. B., Quart. J. Stud. Alc., 25, 498 (1964).
132. Machata, G., Mikrochim. Acta, 262 (1964).
133. Goldbaum, L. R., Domanski, T. J., and Schloegel, E. L., J. Forensic Sci., 9, 63 (1964).

134. Natelson, S. and Stellate, R. L., Clin. Chem., 11, 805 (1965).
135. Natelson, S. and Stellate, R. L., Michrochem. J., 9, 245 (1965).
136. Leibrand, R. J., F&M Facts and Methods, Vol. 6, No. 4, November-December 1965.
137. Machata, G., Oesterr. Chem.-Ztg., 66, 1 (1965).
138. Wallace, J. E. and Dahl, E. V., Amer. J. Clin. Pathol., 46, 152 (1966).
139. Wallace, J. E. and Dahl, E. V., Tech. Bull. Reg. Med. Tech., 36, 152 (1966).
140. Machata, G., Fortschritte de Chemischen Forschung, 6, 4 (1966).
141. Machata, G., Blutalkohol, 4, 252 (1967).
142. Savory, J., Sunderman, F. W., Jr., Roszel, N. O., and Mushak, P., Clin. Chem., 14, 132 (1968).
143. Basette, R. and Glendenning, B. L., Microchem. J., 13, 374 (1968).
144. Jentzsch, D., Kruger, H., Lebrecht, G., Dencks, G., and Gut, J., Z. Anal. Chem., 236, 96 (1968).
145. Hauck, G. and Terfloth, H. P., Chromatographia, 2, 309 (1969).
146. Glendenning, B. L. and Harvey, R. A., J. Forensic Sci., 14, 136 (1969).
147. Gessner, P. K., Anal. Biochem., 38, 499 (1970).
148. Luckey, M. J., J. Forensic Sci., 16, 120 (1971).
149. Coldwell, B. B., Solomonraj, G., Trenholm, H. L., and Wiberg, G. S., Clin. Toxicol., 4, 99 (1971).
150. Machata, G., Perkin-Elmer Clin. Chem. Newsl., Vol. 4, No. 2, pp. 29-32, Winter 1972.
151. Kolb, B., C. Z. Chem. Tech., 1, 87 (1972).
152. Naresh, C. J. and Cravey, R. H., J. Chromatogr. Sci., 10, 263 (1972).
153. Vitenberg, A. G., Ioffe, B. V., and Borisov, V. N., Chromatographia, 7, 610 (1974).
154. Harger, R. N., Raney, B. B., Bridwell, E. G., and Kitchel, M. F., J. Biol. Chem., 183, 197 (1950).
155. Savory, J. and Kaplan, A., Clin. Chem., 12, 559 (1966).
156. Blackmore, D. J., J. Forensic Sci., 8, 73 (1968).
157. Stromme, J. H., Biochem. J., 92, 25P (1964).
158. Fischer, R. and Brantner, H., Naturwissenschaften, 50, 551 (1963).
159. Stromme, J. H., Biochem. Pharmacol., 15, 287 (1966).
160. Merlevede, E. and Peters, J., Arch. Belg. Med. Soc., Hyg., Med. Trav. Med. Leg., 23, 513 (1965).
161. Domar, G., Eredga, A., and Linderholm, H., Acta Chem. Scand., 3, 1441 (1949).
162. Eldjarn, L., Scand. J. Clin. Lab. Invest., 2, 202 (1950).
163. Linderholm, H. and Berg, K., Scand. J. Clin. Lab. Invest., 3, 96 (1951).

164. Gessner, T. and Jakobowski, M., Biochem. Pharmacol., 21, 219 (1972).
165. Kaslander, J., Biochem. Biophys. Acta, 71, 730 (1963).
166. Wells, J. and Koves, E., J. Chromatogr., 92, 442 (1974).
167. Gelder, M. G. and Edwards, G., Brit. J. Psychiatr., 114, 473 (1968).
168. Rollo, I. M., in L. S. Goodman and A. Gilman (Eds.), The Pharmacological Basis of Therapeutics, 4th ed., Macmillan, New York, 1970.
169. Midha, K. K., McGilveray, I. J., and Cooper, J. K., J. Chromatogr., 87, 491 (1973).
170. Robinson, A. B., Partridge, D., Turner, M., Teranishi, R., and Pauling, L., J. Chromatogr., 85, 19 (1973).
171. Pauling, L., Science, 160, 265 (1968).
172. Robinson, A. B. and Pauling, L., Orthomolecular Psychiatry, Freeman, San Francisco, 1973.
173. Thompson, W. E., McKee, H. C., Wehrli, H., and Williams, R., in Anonymous, Chem. & Eng. News, March 14, 1960, pp. 36-37.
174. Jansson, B. O. and Larsson, B. T., J. Lab. Clin. Med., 74, 961 (1969).
175. MacKay, D. A. M., Lang, D. A., and Berdick, M., Drug Cosmetic Ind., 86, 46 (1960).
176. MacKay, D. A. M., Lang, D. A., and Berdick, M., Drug Cosmetic Ind., 86, 166 (1960).
177. Stewart, R. D., Gay, H. H., Erley, D. S., Hake, C. L., and Peterson, J. E., Ind. Hyg. J., 23, 167 (1962).
178. Eriksen, S. P. and Kulkarni, A. B., Science, 141, 639 (1963).
179. Stewart, R. D., Swank, J. D., Roberts, C. B., and Dodd, H. C., Nature, 198, 696 (1963).
180. Larsson, B. T., Acta Chem. Scand., 19, 159 (1965).
181. Levey, S. and Balchum, O. J., J. Lab. Clin. Med., 63, 574 (1964).
182. Richter, V. J. and Tonzetich, J., Arch. Oral Biol., 9, 47 (1964).
183. Rooth, G. and Ostenson, S., Lancet, 2, 1102 (1966).
184. Trotter, M. D., Sulway, M. J., and Trotter, E., Clin. Chim. Acta, 35, 137 (1971).
185. Teranishi, R., Mon, T. R., Robinson, A. B., Cary, P., and Pauling, L., Anal. Chem., 44, 18 (1972).
186. Zlatkis, A., Lichtenstein, H. A., and Tishbee, A., Chromatographia, 6, 67 (1973).
187. Hartman, C. H., Penton, J. R., and Forrester, M., paper presented at 21st Meeting of AAFS, Chicago, February 1969.
188. Morris, R. L., Van Note, W., and Hausler, W. J., Health Lab. Sci., 8, 3, 142 (1971).

189. Biasotti, A. A. and Bradford, L. W., J. Forensic Sci., 9, 65 (1969).
190. Lilijestrand, G. and Linde, P., Scand. Arch. Physiol., 60, 273 (1930).
191. Harger, R. N., Lamb, E. B., and Hulpieu, R. R., J. Amer. Med. Assoc., 110, 779 (1938).
192. Harger, R. N. and Forney, R. B., paper presented at 5th International Meeting of Alcohol and Road Traffic, Freiberg, Germany, 1969.
193. Spector, N. H., Science, 172, 57 (1971).
194. Payne, J. P., Europa Med., 1 (March 1967).
195. Milner, G. and Landauer, A. A., Med. J. Aust., 1, 1280 (1971).
196. Kitagawa, T. and Wright, B. M., Brit. Med. J., 2, 652 (1962).
197. Yamamoto, K. and Ueda, N., Forensic Sci., 1, 207 (1972).
198. Brokenstein, R. F., Breath Tests to Determine Alcoholic Influence, Ind. Ed., Indiana State Police, Indianapolis, Ind., 1957.
199. Borkenstein, R. F. and Smith, H. W., Medicine, Science and the Law, 1, 13 (1961).
200. Forrester, G. C., Alcohol, Traffic Accidents and Chemical Test Evidence: The Photoelectric Intoximeter, Intoximeter Ass., Niagara Falls, N.Y., 1960.
201. Monnier, D. and Etzlinger, L., Rev. Int. Criminol. Police Tech., 25, 305 (1971).
202. Bay, H. W., Blurton, K. F., Lieb, H. C., and Oswin, H. G., Nature, 240, 52 (1972).
203. Jones, T. P. and Williams, P. M., in Abstracts of the 6th International Meeting of Forensic Sciences, Edinburgh, September 1972.
204. Borg-Warner ALERT™, Model J2A-1000, Borg-Warner Corp., Des Plaines, Ill.
205. Burroughs, J. E., Nowak, J. A., and Hoppesch, J. P., Chem. Abstr., 79(15), 88110a (1973).
206. Harte, R. A., J. Forensic Sci., 16, 4, 493 (1971).
207. Kamei, H. and Koezuka, J., Chem. Abstr., 79, 14256f (1973).
208. Kamei, H., Koezuka, J., and Imooka, M., Chem. Abstr., 80, 651q (1974).
209. Collier, C. R., J. Appl. Physiol., 9, 25 (1956).
210. Levey, S., Balchum, O. J., Medrano, V., and Jung, R., J. Lab. Clin. Med., 63, 574 (1964).
211. Sulway, M. J. and Malins, J. M., Lancet, 2, 736 (1970).
212. Sulway, M. J., Trotter, E., Trotter, M. D., and Malins, J. M., Postgrad. Med. J., Suppl., June 1971, p. 382.
213. Teranishi, R., Buttery, R. G., and Lundin, R. E., Anal. Chem., 34, 1033 (1962).
214. Teranishi, R. and Mon, T. R., Anal. Chem., 36, 1490 (1964).
215. Flath, R. A., Forrey, R. R., and Teranishi, R., J. Food Sci., 34, 382 (1969).

216. Horning, E. C., Vanden Heuvel, W. J. A., and Creech, B. G., in D. Glick (Ed.), Methods of Biochemical Analysis, Vol. 11, Interscience, New York, 1963.
217. Horning, M. G., in Biochemical Applications of Gas Chromatography, Vol. 2, Plenum Press, New York, 1968, pp. 53-86.
218. Novotny, M. and Zlatkis, A., Chromatogr. Rev., 14, 1 (1971).
219. Zlatkis, A. and Liebich, H. M., Clin. Chem., 17, 592 (1971).
220. Field, T. G., Orynich, R. E., and Gilbert, J. B., 150th ACS Meeting, Div. Biol. Chem., Atlantic City, N.J., September 1965.
221. Van Haaften, A. B. and Sie, S. T., Amer. Ind. Hyg. Ass. J., 26, 52 (1965).
222. Zlatkis, A., Bertsch, W., Lichtenstein, H. A., Tishbee, A., Shunbo, F., Liebich, H. M., Coscia, A. M., and Fleischer, N., Anal. Chem., 45, 763 (1973).
223. Zlatkis, A., Lichtenstein, H. A., Tishbee, A., Bertsch, W., Shunbo, F., and Liebich, H. M., J. Chromatogr. Sci., 11, 299 (1973).
224. Horning, E. C. and Horning, M. G., in R. E. Olson (Ed.), Methods in Medical Research, Year Book Medical Publishers, Chicago, 1970.
225. Jellum, E., Stokke, O., and Eldjarn, L., Scand. J. Clin. Lab. Invest., 27, 273 (1971).
226. Karoum, F., Ruthven, C. R. J., and Sandler, M., Clin. Chim. Acta, 20, 427 (1968).
227. Witten, T. A., Levine, S. P., Killian, M. T., Boyle, P. J. R., and Markey, S. P., Clin. Chem., 19, 963 (1973).
228. Matsumoto, K. E., Partridge, D. H., Robinson, A. B., Pauling, L., Flath, R. A., Mon, T. R., and Teranishi, R., J. Chromatogr., 85, 31 (1973).
229. Pauling, L., Robinson, A. B., Teranishi, R., and Cary, P., Proc. Nat. Acad. U. S., 68, 2374 (1971).
230. Gibbs, H. D., J. Biol. Chem., 72, 649 (1927).
231. Theis, R. C. and Benedict, S. R., J. Biol. Chem., 61, 67 (1924).
232. Hammond, K. B. and Goodman, S. I., Clin. Chem., 16, 212 (1970).
233. Witten, T. A., Levine, S. P., King, J. O., and Markey, S. P., Clin. Chem., 19, 586 (1973).
234. Levine, S. P., Naylor, J. L., and Pearce, J. B., Anal. Chem., 45, 1560 (1973).
235. Williamson, J. R., Scholz, R., Browning, E. T., Thurman, R. G., and Fukami, M. H., J. Biol. Chem., 244, 5044 (1969).
236. Lundquist, F., in Y. Israel and J. Mardones (Eds.), Biological Basis of Alcoholism, Wiley-Interscience, New York, 1971.
237. Davis, V. E. and Walsh, M. J., in Y. Israel and J. Mardones (Eds.), Biological Basis of Alcoholism, Wiley-Interscience, New York, 1971.

238. Davis, V. E., Walsh, M. J., and Yamanaka, Y., J. Pharmacol. Exp. Ther., 174, 401 (1970).
239. Kalent, H., Hawkins, R. D., and Czaja, C., Amer. J. Physiol., 204, 849 (1963).
240. Merry, J. and Marks, V., in G. A. Martini and C. Bode (Eds.), Changes Induced by Alcohol, Springer-Verlag, New York, 1971.
241. Schultz, T. H., Flath, R. A., and Mon, T. R., J. Agr. Food Chem., 19, 1060 (1971).
242. Cornu, A. and Massot, R., Compilation of Mass Spectral Data, Heyden & Son, London, 1966.
243. Cornu, A. and Massot, R., First Supplement to Compilation of Mass Spectral Data, Heyden & Son, London, 1967.
244. Hill, D. W. and Butler, R. A., Nature, 189, 488 (1961).
245. Lester, D., Anal. Chem., 36, 1810 (1964).
246. Northington, M. S. and Owens, G., J. Gas Chromatogr., 3, 107 (1965).
247. Baker, R. N., Alenty, A. L., and Zack, J. F., Jr., J. Chromatogr. Sci., 7, 312 (1969).
248. Payne, J. P., Foster, D. V., Hill, D. W., and Wood, D. G. L., Brit. Med. J., 3, 819 (1967).
249. LeBlanc, A. E., Can. J. Physiol. Pharmacol., 21, 275 (1968).
250. Basette, R., Ozeris, S., and Whitnah, C. H., Anal. Chem., 34, 1540 (1962).
251. Stoner, E., Cowburn, D., and Craig, L. C., Anal. Chem., 47, 344 (1975).
252. Virtanen, A. I. and Pulkki, L., J. Amer. Chem. Soc., 50, 3138 (1928).
253. Dybing, F. and Dybing, O., Acta Pharmacol., 1, 270 (1945).

AUTHOR INDEX

Numbers in parentheses are reference numbers and indicate that an author's work is referred to although his name is not cited in the text. Underlined numbers give the page on which the complete reference is listed.

A

Adlard, E. R., 1, 38(5), 39, 40, 46, 75
Adler, N., 3(13), 63(13), 64, 75
Adriani, J., 38(75), 42, 78
Aebi, H., 89(28), 181
Ahtee, L., 96, 182
Albin, M. S., 38(113), 48(113), 79
Alenty, A. L., 171(247), 177(247), 189
Allain, P., 3(15), 69(15), 70(15), 71(15), 75
Alper, M. H., 38(90), 47(90), 51(90), 54(90), 78
Alvares, A. P., 89(39), 181
Ariyoshi, T., 89(14), 180
Armstrong, D. T., 89(32), 90(32), 181
Aronow, L., 83
Asao, Y., 1(2), 75
Assimos, A., 112(104), 120, 171(104), 183
Aul, F., 20(60), 37(60), 38, 77
Ayres, S. M., 20(57), 28(57), 33, 35, 77

B

Bacchin, P., 89(35), 181
Bacon, V. A., 133(129), 184
Bagwell, E. E., 38(97), 38(105), 51(105), 57(97), 78, 79
Baird, W. W., 112(116), 126, 161(116), 184
Baker, J. M., 38(76), 41, 78
Baker, R. N., 171(247), 177, 189
Balchum, O. J., 154(181), 157(210), 186, 187
Bannard, R. A. B., 3(16), 74(16), 76
Bard, J., 101(67), 102, 171(67), 182
Barnard, G. P., 82
Barrais, C., 105(79), 183
Barrett, M. J., 109, 183
Basette, R., 134(143), 140, 171(143), 171(250), 185, 189
Bassler, G. C., 82
Bates, D. V., 10(30), 76
Bay, H. W., 155(202), 187
Beckey, H. D., 82
Beckham, L. M., 38(83), 44, 45, 51(83), 78
Beckman, H., 69, 80
Belleville, J. W., 38(84), 38(89),

48(84), 48(89), 49(89), 50(89), 78
Benedict, S. R., 162(231), 188
Berchild, K. M., 112(115), 129(115), 131(115), 184
Berck, B., 66, 67, 68, 69, 80
Berdick, M., 154(175), 154(176), 186
Berg, K., 150(163), 185
Bertsch, W., 161(222), 161(223), 165(222), 165(223), 168(222), 168(223), 169(223), 170(223), 188
Bevenue, A., 69, 80
Beynon, J. H., 81, 82
Biasotti, A. A., 154(189), 187
Biemann, K., 81
Blackmore, D. J., 20(59), 77, 140, 185
Blume, P., 112(115), 129, 131, 184
Blumenfeld, C. M., 48(130), 80
Blurton, K. F., 155(202), 187
Bode, C., 189
Boer, H., 5, 76
Bonelli, E. J., 126, 184
Bonnichsen, R., 105(81), 112(99), 112(113), 118, 128, 129, 161, 161(99), 162, 183, 184
Borisov, V. N., 134(153), 148(153), 185
Boyle, P. J. R., 161(227), 168(227), 173(227), 188
Brachet-Liermain, A., 38(94), 51(94), 54, 55, 78
Bradford, L. W., 154(189), 187
Brahm-Vogelsanger, A., 112(102), 183
Brantner, H., 150(158), 185
Brewer, H. W., 38(85), 38(86), 38(89), 48(85), 48(86), 48(89), 49, 49(85), 49(89), 50(89), 51, 51(85), 54, 78
Bridges, J. C., 133(129), 184
Bridwell, E. G., 134(154), 185
Brokenstein, R. F., 155(198), 155(199), 187
Brown, G. A., 112(118), 112(119), 131, 133, 171(118), 171(119), 184
Browning, E. T., 170(235), 188
Budzikiewicz, H., 38(84), 48(84), 78, 81, 82
Bunker, J. P., 48(130), 80
Burchfield, H. P., 111(89), 183
Burnham, A. K., 20(46), 76
Burroughs, J. E., 155(205), 187
Butler, R. A., 1(4), 2(4), 38(4), 38(87), 38(103), 38(122), 39, 48(87), 49, 51(4), 51(103), 51(122), 75, 78, 79, 171(244), 175, 189
Buttery, R. G., 158(213), 187

C

Cadman, W. J., 41, 79, 101, 101(59), 101(66), 102, 103, 104, 111, 120, 135, 171(66), 182
Cailleux, A., 3(15), 69(15), 70(15), 71(15), 75
Calas, C., 31(70), 77
Calder, G. V., 20(46), 76, 96, 97, 98, 99, 100, 101, 182
Campbell, M. M., 82
Carnes, M. A., 38(99), 78
Caroff, J., 38(94), 51(94), 54, 55(94), 78
Carr, Z., 20(58), 28(58), 33(58), 35, 77
Carroll, D. G., 10(31), 76
Carter, E. A., 89(24), 181
Cary, P., 154(185), 158(185), 159(185), 160(185), 161(185), 161(229), 165(185), 170(185), 186, 188
Casselman, A. A., 3(16), 74, 76
Cawley, L. P., 112(115), 129(115), 131(115), 184

Cazes, J., 81
Chambliss, K. W., 20(61), 33, 77
Chang, R. C., 20, 21, 77
Chapman, J., 38(116), 48(116), 50, 52, 79
Cheek, C. H., 2(7), 75
Cheneau, M., 38(78), 41, 78
Chenoweth, M. B., 38(86), 38(106), 47, 48(86), 57, 78, 79, 80
Ching, N. P., 20(62), 77
Chiou, W. L., 3(11), 3(12), 38(11), 38(12), 59, 61, 75, 80
Christenson, H. E., 20(64), 36(64), 77
Chundela, B., 112(93), 112(94), 115, 183
Ciccarelli, H. E., 38(115), 51(115), 79
Clarke, E. G. C., 128, 184
Clayton, J. W., 49, 80
Coburn, T. C., 133(129), 184
Cohen, E. N., 38(84), 38(89), 48, 48(84), 48(85), 49, 50, 51, 51(85), 54, 78
Cohen, J. E., 10(31), 76
Coldwell, B. B., 112(100), 120, 134(149), 143, 161(100), 171(100), 183, 185
Colehour, J. K., 112(108), 122, 123, 171(108), 184
Collier, C. R., 157, 187
Collins, A., 89(19), 180
Conney, A. H., 89(11), 180
Conolly, M. E., 59(136), 80
Cooper, J. D. H., 112(109), 123, 184
Cooper, J. K., 152, 186
Cornu, A., 171, 171(242), 171(243), 189
Coscia, A. M., 161(222), 165(222), 168(222), 188
Coumbis, R. J., 112(105), 183
Cowburn, D., 171(251), 177, 179(251), 189
Craig, L. C., 171(251), 177, 179(251), 189
Cravey, R. H., 101, 101(60), 134(152), 155, 182, 185
Creaven, P. J., 89(26), 101(82), 105, 106, 181, 183
Creech, B. G., 159, 188
Creswell, C. J., 82
Cudmore, J. W., 1(3), 38(3), 41, 75
Cullen, T. E., 69, 80
Curry, A. S., 3(9), 38(9), 48, 75, 105(80), 112(106), 120, 131, 134(130), 171(80), 171(106), 171(130), 175, 183, 184
Czaja, C., 170(239), 189
Czubryt, J. J., 20(45), 76

D

Dagleish, C. E., 86(7), 161(7), 164, 171(7), 180
Dahl, E. V., 134(138), 134(139), 137, 140, 141, 161(139), 171(139), 185
Dal Nogare, S., 80
Dave, S. B., 20(41), 76
Davies, D. D., 38(117), 54, 79
Davies, D. S., 59(136), 80
Davis, R. A., 101(68), 171(68), 182
Davis, V. E., 91(50), 92, 92(52), 170(237), 170(238), 182, 188, 189
De Bruijn, J. W., 57, 80
De Carli, L. M., 89(43), 89(44), 89(45), 89(46), 89(47), 89(49), 90(49), 181, 182
De Graff, A. C., Jr., 7, 8, 9, 9(25), 10, 76
De Matteis, F., 89(18), 180
Dencks, G., 134(144), 185
Desty, D. H., 76

Dewitt, H., 38(99), 78
Diamond, I., 101(72), 107(72), 108(72), 109(72), 182
Dillon, R. T., 29(68), 77
Djerassi, C., 81, 82
Dodd, H. C., 38(93), 54(93), 78, 154(179), 186
Doerge, R. F., 83
Dollery, C. T., 59, 80
Domanski, T. J., 134, 134(133), 136(133), 140(133), 161(133), 171(133), 184
Domar, G., 150(161), 185
Dominguez, A. M., 20(64), 36, 36(73), 38(73), 77
Doty, V., 20(49), 21(49), 23(49), 77
Douglas, R., 38(119), 51(119), 54, 79
Downs, S., 38(74), 39, 77
Draffan, G. H., 59(136), 80
Drake, N. L., 3(20), 76
Drawert, F., 112(95), 112(96), 115, 117, 161(95), 183, 184
Du Bois, A. B., 10(32), 76
Dubowski, K. M., 101, 154(65), 182
Duffield, A. M., 112(120), 133(120), 135(120), 161(120), 184
Duncan, W. A. M., 53(133), 80
Duritz, G., 107, 134(131), 140, 141, 171(83), 171(131), 183, 184
Dybing, F., 177, 189
Dybing, O., 177, 189
Dyfverman, A., 38(104), 51(104), 79

E

Edwards, A. W. T., 20(53), 29(53), 30(53), 77
Edwards, G., 152, 186
Eldjarn, L., 150(162), 161(225), 185, 188
Ellis, G. P., 83
Epstein, R. M., 38(92), 38(126), 53, 78, 79
Eredga, A., 150(161), 185
Eriksen, S. P., 154(178), 186
Erley, D. S., 57(134), 80, 154(177), 186
Ettre, L. S., 80, 81, 82
Etzlinger, L., 155(201), 187

F

Fabian, L. W., 38(99), 78
Farhi, L. E., 20(53), 29, 30, 31, 77
Farlow, R., 86(6), 161(6), 171(6), 174(6), 180
Felgenhauer, R., 112(95), 115(95), 161(95), 183
Feneslau, C., 82
Ferrus, L., 38(94), 51(94), 54, 55(94), 78
Field, F. H., 82
Field, T. G., 161(220), 162, 188
Fink, B. R., 38(125), 51(125), 57, 78, 79
Finkle, B. S., 112(111), 124, 161(111), 171(111), 184
Fischer, R., 150(158), 185
Fisher, T. L., 3(18), 71(18), 76
Flacke, W., 38(90), 47(90), 51(90), 54(90), 78
Flath, R. A., 159(215), 161(228), 171(228), 171(241), 187, 188, 189
Fleischer, N., 161(222), 165(222), 168(222), 188
Fontan, C. R., 112(98), 117(98), 118(98), 119(98), 161(98), 171(98), 183
Forbes, J., 38(110), 47(110), 79

Foris, A., 75(145), 80
Forney, R. B., 85(3), 102, 154(192), 180, 182, 187
Forrester, G. C., 155(200), 187
Forrester, M., 154(187), 186
Forrey, R. R., 159(215), 187
Forster, R. E., 10, 10(34), 76
Foster, D. V., 171(248), 189
Foster, N. G., 81
Fowler, W. S., 10(30), 76
Fox, J. E., 110, 110(85), 111, 161(85), 171(85), 183
Fraccaro, C., 38(118), 48(118), 79
Franklin, J. L., 82
Freeman, J., 38(103), 51(103), 78
Freudiger, J. B., 112(103), 183
Freund, G., 112(107), 121, 161(107), 171(107), 184
Fritz, J. S., 20, 20(46), 21, 76, 77
Fukami, M. H., 170(235), 188

G

Gadsen, R. H., 38(82), 38(97), 38(105), 38(107), 44, 46, 47, 51(82), 51(105), 57, 78, 79, 110(86), 110(87), 111, 111(87), 112, 113, 161(86), 161(87), 171(86), 171(87), 183
Galla, S. J., 20(50), 24, 25, 26, 27, 77
Gambaretto, G., 38(118), 48(118), 79
Gang, H., 89(35), 89(42), 89(46), 89(48), 91(42), 92(42), 181, 182
Garattini, S., 83
Gay, H. H., 154(177), 186
Gelder, M. G., 152, 186
Gerarde, H. W., 103, 182
Gesser, H. D., 20(45), 76
Gessner, P. K., 134(147), 141, 142, 143, 171(147), 185
Gessner, T., 150, 150(164), 186
Gibaldi, M., 83
Gibbs, H. D., 162(230), 188
Giddings, J. C., 81
Gigon, P. L., 89(29), 181
Gilbert, J. B., 161(220), 162(220), 188
Gillette, J. R., 89(29), 89(30), 89(31), 181
Gilman, A., 83, 180, 186
Gisvold, O., 83
Gjaldbaek, J. C., 38(101), 38(114), 48(114), 78, 79
Glendenning, B. L., 134(143), 134(146), 140, 171(143), 171(146), 185
Glick, D., 188
Gloesner, E., 38(100), 78
Gohlke, R., 57(134), 80
Goldbaum, L. R., 20(64), 36(64), 36(73), 38(73), 77, 134, 134(133), 136, 140, 161(133), 171(133), 184
Goldberg, L., 104, 183
Goldfien, A., 83
Goldstein, A., 83, 92, 182
Goodman, L. S., 83, 180, 186
Goodman, S. I., 169, 188
Gossmann, K., 38(110), 47(110), 79
Goth, A., 83, 85, 86(1), 152(1), 180
Gottlieb, L. S., 89(33), 181
Gram, T. E., 89(29), 181
Grant, G. L., 112(100), 120, 161(100), 171(100), 183
Griffiths, W. C., 101(72), 107, 108, 109, 182
Grollman, A., 10(33), 76
Grushka, E., 81
Gudzinowicz, B. J., 1(6), 2, 75, 80, 81, 82
Gudzinowicz, M. J., 80, 81, 82

Guichon, G., 38(78), 41, 78
Guild, L. V., 20(60), 37, 38, 77
Gumoei, U., 89(15), 180
Gut, J., 134(144), 185

H

Habgood, H. W., 81
Hagler, K., 3(10), 38(10), 59, 75
Hake, C. L., 154(177), 186
Hall, K. D., 38(74), 39, 77
Halpern, B., 133(129), 184
Hamilton, L. H., 10, 10(27), 10(28), 10(29), 11, 15, 20(27), 24, 27, 76
Hammarstrand, K., 18, 19, 76
Hamming, M. C., 81
Hammond, K. B., 169, 188
Harger, R. N., 96(58), 102, 134, 134(154), 154(191), 154(192), 182, 185, 186
Harley, J., 39(127), 79
Harris, R. J., 115(121), 184
Harris, W. E., 81
Harte, R. A., 155(206), 187
Hartigan, M. J., 112(117), 131, 132, 161(117), 184
Hartman, C. H., 154(187), 186
Hartree, E. F., 89(25), 181
Harvey, R. A., 134(146), 140, 154, 171(146), 185
Hauck, G., 134(145), 140, 185
Hausler, W. J., 154(188), 155(188), 186
Hawkins, R. D., 170(239), 189
Hayes, W. V., 20(40), 76
Hedjfall, B., 112(113), 128(113), 184
Hedley-Whyte, J., 38(124), 51(124), 79
Herban, J., 3(18), 71(18), 76
Herbert, P., 38(111), 38(120), 46(111), 51(111), 51(120), 79
Hessel, D. W., 112(101), 171(101), 183
Hessler, O., 38(110), 47(110), 79
Higgins, E., 20(49), 21(49), 23(49), 77
Higuchi, T., 4, 76
Hill, D. W., 1, 1(1), 1(4), 2(4), 4, 5, 6, 38(4), 38(5), 38(119), 39, 40, 46, 51(4), 51(119), 54(119), 75, 79, 171(244), 171(248), 175, 189
Hill, E., 89(17), 180
Hill, H. C., 82
Hill, R., 38(116), 48(116), 50(116), 52(116), 79
Hirsch, R. F., 20(43), 76
Hitomi, Y., 38(123), 51(123), 79
Hoberecht, H. D., 112(117), 131(117), 132(117), 161(117), 184
Hollis, O. L., 20, 20(40), 20(42), 76
Holtz, P., 91, 182
Homma, T., 20(53), 29(53), 30(53), 77
Hoppesch, J. P., 155(205), 187
Horning, E. C., 86(7), 86(8), 159, 161(7), 161(8), 161(224), 163, 163(7), 163(8), 164, 164(8), 165, 166, 167, 171, 171(7), 171(8), 180, 188
Horning, M. G., 86(7), 86(8), 159, 161(7), 161(81), 161(224), 163, 163(7), 163(8), 164, 164(8), 165, 166, 167, 171, 171(7), 171(8), 180, 188
Horrocks, L. A., 38(113), 48(113), 79
Hsiao, J. H., 3(12), 38(12), 59, 75
Hubner, F. N., 20(43), 76
Hughes, F. W., 85(3), 180
Hulpieu, R. R., 154(191), 187
Hurst, G., 3(9), 38(9), 48(9), 75, 134(130), 171(130), 175(130), 184
Hutterer, F., 89(34), 181

I

Iber, F. L., 89(22), 181
Ikels, K. G., 20(54), 29, 31, 77
Imooka, M., 155(208), 187
Ioffe, B. V., 134(153), 148(153), 185
Iseri, O. A., 89(33), 181
Ishii, H., 89(41), 181
Israel, Y., 188
Isselbacher, K. J., 89(24), 181

J

Jacobs, E. S., 38(96), 56, 57, 59, 78
Jain, N. C., 101, 112(110), 124, 130, 154, 155, 171(110), 177, 178, 182, 184
Jakobowski, M., 150, 150(164), 186
Janak, J., 112(93), 112(94), 115, 183
Jansson, B. O., 154(174), 155, 157, 186
Jascot, M. J., 3(18), 71(18), 76
Jawetz, E., 83
Jay, B. E., 10, 11, 12, 13, 14, 20(26), 20(49), 20(65), 21(26), 21(49), 21(65), 23(49), 24, 76, 77
Jellum, E., 161(225), 188
Jentzsch, D., 134(144), 185
Johannsmeir, K., 101(70), 182
Johns, T., 20(63), 41, 77, 79, 101, 101(59), 101(66), 102, 103, 104, 111, 120, 135, 171(66), 182
Johnson, R. L., 5(25), 7(25), 8(25), 9(25), 10(25), 76
Johnson, S. L., 49(131), 80
Johnstone, R. A. W., 81
Jones, A. L., 89(32), 90(32), 181
Jones, R. A., 81
Jones, T. P., 155(203), 187
Joubert, S. M., 89(16), 89(20), 180
Judson, B. A., 92, 182
Jung, R., 157(210), 187
Junk, G. A., 20(46), 76
Juvet, R. S., Jr., 80

K

Kaiser, R., 80
Kalent, H., 170(239), 189
Kalman, S. M., 83
Kambarat, T., 20(44), 76
Kamei, H., 155(207), 155(208), 187
Kaplan, A., 138, 185
Karasek, F. W., 82
Karoum, F., 161(226), 188
Kaslander, J., 150(165), 186
Kater, R. M. H., 89(22), 181
Keillin, D., 89(25), 181
Keller, R. A., 81
Kelly, A. B., 38(122), 51(122), 79
Kent, N. R., 3(9), 38(9), 48(9), 75, 134(130), 171(130), 175(130), 184
Keulemans, A. I. M., 80
Kiku, N., 89(15), 180
Killian, M. T., 161(227), 168(227), 173(227), 188
King, B. D., 38(76), 41, 78
King, J. O., 170(233), 188
Kirk, P. L., 112(98), 117(98), 118(98), 119(98), 161(98), 171(98), 183
Kiser, R. W., 82
Kitagawa, T., 155(196), 187
Kitchel, M. F., 134(154), 185
Klocke, F. J., 29, 77
Knox, K. L., 86(7), 161(7), 163(7), 171(7), 180
Koezuka, J., 155(207), 155(208), 187
Kolb, B., 134(151), 185
Kory, R. C., 10(28), 10(29), 76
Kosaka, F., 38(123), 57(123), 79

Koss, R., 59, 80
Koudela, S., 38(109), 79
Koves, E., 150(166), 152, 186
Kowblansky, M., 20(43), 76
Krantz, J. C., 49(131), 80
Kretchmer, H. E., 38(113), 48(113), 79
Kruger, H., 134(144), 185
Kulkarni, A. B., 154(178), 186
Kuntzman, R., 89(39), 181
Kupfer, G., 112(95), 112(96), 115(95), 117, 161(95), 183, 184

L

Laasberg, L. H., 38(124), 51(124), 79
Ladd, S. L., 128, 184
La Du, B. N., 83
Laessig, R. H., 101(74), 103, 104, 105, 106, 171(74), 182
Lamb, E. B., 154(191), 187
Landauer, A. A., 154, 187
Lang, D. A., 154(175), 154(176), 186
Larson, E. R., 38(106), 47, 79
Larsson, B. T., 154(174), 154(180), 155, 157, 186
Law, M. R., 112(114), 126(114), 128(114), 129(114), 171(114), 184
Leathard, D. A., 81
Le Blanc, A. E., 171(249), 189
Lebrecht, G., 134(144), 185
Lederberg, J., 133(129), 184
Lee, G. de J., 10(32), 76
Lee, J., 20(58), 28(58), 33(58), 77
Lee, M. L., 86(6), 161(6), 171(6), 174(6), 180
Lefevre, A., 89(41), 181
Legendre, G., 105(79), 183
Lehman, J. G., 75(145), 80
Leibrand, R. J., 134(136), 185
Leighton, D., 31(70), 77
Lesnini, D. G., 3, 4(22), 76
Lester, D., 171(245), 175, 176, 189
Levey, S., 154(181), 157, 186, 187
Levin, W., 89(39), 181
Levine, R. R., 83
Levine, S. P., 161(227), 168(227), 170(233), 170(234), 173(227), 188
Levinthal, E. C., 133(129), 184
Lichtenstein, H. A., 154(186), 159(186), 161(186), 161(222), 161(223), 165(222), 165(223), 168(186), 168(222), 168(223), 169(223), 170(223), 171(186), 186, 188
Lieb, H. C., 155(202), 187
Lieber, C. S., 85(2), 89, 89(33), 89(34), 89(35), 89(36), 89(37), 89(38), 89(39), 89(40), 89(41), 89(42), 89(43), 89(44), 89(45), 89(46), 89(47), 89(48), 89(49), 90, 90(49), 91(42), 92(42), 180, 181, 182
Liebich, H. M., 159(219), 161(219), 161(222), 161(223), 165, 165(219), 165(222), 165(223), 168(222), 168(223), 169(223), 170(223), 188
Lilijestrand, G., 154(190), 187
Linde, H. W., 38(87), 48(87), 49, 78
Linde, P., 154(190), 187
Linderholm, H., 150(161), 150(162), 185
Ling, M. S., 49(131), 80
Linnenbom, V. J., 2(7), 75
Linturi, M., 112(99), 118, 161, 161(99), 162, 183
Littlewood, A. B., 81
Lloyd, M. I., 20(60), 37(60), 38, 77
Lowe, H. J., 3(10), 38(10), 38(81),

38(83), 38(102), 38(108), 42, 43, 44, 45, 46, 51(81), 51(83), 51(108), 59, 75, 78, 79
Lu, G., 49, 80
Luckey, M. J., 134(148), 142, 154(148), 155, 161(148), 185
Lukaci, J., 38(109), 79
Lukas, D. S., 20(57), 28(57), 33, 35, 77
Lundgren, G., 105(81), 183
Lundin, R. E., 158(213), 187
Lundquist, F., 170(236), 188
Lyons, H., 101(67), 102, 171(67), 182

M

Machata, G., 112(97), 117, 134(132), 134(137), 134(140), 134(141), 134(150), 137, 144, 145, 147, 148(150), 149, 171(150), 183, 184, 185
McConnell, M. L., 86(6), 161(6), 171(6), 174(6), 180
McCord, W. M., 38(82), 38(97), 38(107), 44, 46, 47, 51(82), 57(97), 78, 79, 110(86), 110(87), 111, 111(87), 112, 113, 161(86), 161(87), 171(86), 171(87), 183
McFadden, W. H., 82
McGilveray, I. J., 152, 186
MacKay, D. A. M., 154(175), 154(176), 186
McKee, H. C., 154(173), 155(173), 156, 161(173), 186
McKenna, M. J., 112(114), 126(114), 128(114), 129(114), 171(114), 184
McLafferty, F. W., 81
McNair, H. M., 126, 184
Maickel, R. P., 183
Malins, J. M., 158(211), 158(212), 171(211), 171(212), 187
Mandel, H. G., 83
Manning, R. T., 89(13), 180
Mardones, J., 188
Margaria, R., 29(68), 77
Maricq, L., 112(90), 112(91), 114, 114(90), 114(91), 123, 171(90), 171(91), 183
Markey, S. P., 161(227), 168(227), 170(233), 173(227), 188
Marks, V., 170(240), 189
Martin, H. F., 80, 81, 82
Martini, G. A., 189
Marver, H. S., 89(19), 180
Massot, R., 171, 171(242), 171(243), 189
Master, R., 105(79), 183
Mather, A., 112(104), 120, 171(104), 183
Mathias, J. A., 38(117), 54, 79
Matsumoto, K. E., 161(228), 171, 188
Medrano, V., 157(210), 187
Mello, N. K., 89(21), 180
Melmon, K. L., 83
Mendelson, J. H., 89(21), 180
Menning, C., 31(69), 77
Merlevede, E., 150(160), 185
Merritt, C., Jr., 82
Merry, J., 170(240), 189
Meyers, F. H., 83
Midha, K. K., 152, 186
Mikkelsen, L., 112(112), 125(112), 161(112), 184
Milner, G., 154, 187
Misra, P. S., 89(41), 89(42), 89(46), 91(42), 92(42), 181
Modglin, F. R., 112(101), 171(101), 183
Molle, L., 112(90), 112(91), 114, 114(90), 114(91), 123, 171(90), 171(91), 183

Mon, T. R., 154(185), 158(185), 158(214), 159(185), 160(185), 161(185), 161(228), 165(185), 170(185), 171(228), 171(241), 186, 187, 188, 189
Monnier, D., 155(201), 187
Mooi, J., 4(22), 76
Morikawa, K., 38(125), 51(125), 57, 79
Morrelli, H. F., 83
Morris, R. L., 154(188), 155, 186
Muir, J., 38(116), 48(116), 50(116), 52(116), 79
Murray, M. H., 112(114), 126(114), 128(114), 129(114), 171(114), 184
Mushak, P., 134(142), 138(142), 139(142), 171(142), 185

N

Naresh, C. J., 134(152), 185
Nash, J. B., 101(69), 103(69), 171(69), 182
Natelson, S., 20(55), 28(66), 29, 31, 31(69), 31(70), 31(71), 32, 33, 34, 77, 134(134), 134(135), 137, 138, 139, 161(134), 161(135), 171(134), 171(135), 185
Naylor, J. L., 170(234), 188
Nel, W., 39(127), 79
Nelson, T. F., Jr., 20(62), 77
Neylan, D., 112(118), 131(118), 171(118), 184
Niazi, S., 61, 80
Noble, E. P., 96, 182
Noehren, T. H., 1(3), 38(3), 41, 75
Norris, F., 38(74), 39, 77
Northington, M. S., 38(80), 41, 78, 171(246), 176, 189
Notari, R. E., 82
Nouse, D. C., 20(61), 33, 77
Novak, J., 101(71), 182
Novotny, M., 86, 159, 161(6), 171, 174, 180, 188
Nowak, J. A., 155(205), 187

O

O'Connell, A. W., 20(43), 76
Ohta, K., 38(123), 51(123), 79
Ohzeki, K., 20(44), 76
Oleksyk, S. K., 101(72), 107(72), 108(72), 109(72), 182
Olling, M., 38(98), 57(98), 60(98), 78
Olson, R. E., 188
Onchi, Y., 1(2), 75
Orkin, L. R., 38(77), 38(79), 42, 78
Orme-Johnson, W. H., 89(23), 181
Orynich, R. E., 161(220), 162(220), 188
Ostenson, S., 154(183), 186
Osterhaus, F., 101(70), 182
Oswin, H. G., 155(202), 187
Ottenstein, D. M., 20(50), 24, 25, 26, 27, 77
Owens, D. R., 39(128), 79
Owens, G., 38(80), 41, 78, 171(246), 176, 189
Oza, T. M., 3(21), 76
Ozeris, S., 171(250), 189

P

Papenberg, J., 89(28), 181
Parker, K. D., 112(98), 117, 118, 119, 161(98), 171(98), 183
Partridge, D. H., 153(170), 161(170), 161(228), 170(170), 171(228), 186, 188
Paterson, J. W., 59, 80
Patterson, D. A., 101, 182
Patzelova, V., 20, 38(38), 47(38), 76

Pauling, L., 153(170), 154(171), 154(172), 154(185), 158(185), 159(185), 160(185), 161(170), 161(185), 161(228), 161(229), 165(185), 170(170), 170(185), 171(228), 186, 188
Payne, J. P., 154, 171(248), 187, 189
Pearce, J. B., 170(234), 188
Penton, J. R., 154(187), 186
Pereira, W. E., 112(120), 133, 135, 161(120), 184
Perl, W., 38(126), 79
Perrier, D., 83
Peters, J., 150(160), 185
Peters, J. P., 35(72), 77
Peterson, J. E., 154(177), 186
Phillips, O. C., 38(88), 48(88), 49, 78
Phillips, T. R., 38(128), 79
Pingree, H., 20(49), 21(49), 23(49), 77
Porter, R., 75, 77
Pottie, R. F., 3(16), 74(16), 76
Powell, H., 3(9), 38(9), 48(9), 75, 134(130), 171(130), 175(130), 184
Premel-Cabic, A., 3(15), 69, 70, 71, 75
Pretorius, V., 39(127), 79
Pulkki, L., 175(252), 189
Purcell, J. E., 112(117), 131(117), 132(117), 161(117), 184
Purchase, I. F. H., 2(8), 38(8), 48, 51(8), 75
Purnell, J. H., 81
Purnell, H., 182

R

Rackow, H., 38(92), 38(126), 53(92), 78, 79
Rahn, H., 29, 77
Ramsey, L. H., 20(56), 33, 34, 35, 77
Raney, B. B., 134(154), 185
Rauwns, A. G., 38(98), 57, 60, 78
Raventos, J., 53(133), 80
Reed, D., 101, 101(60), 182
Reed, R. I., 81
Reese, W. N., 89(26), 181
Rehder, K., 38(110), 47(110), 79
Reichigl, M., Jr., 89(19), 180
Remmer, H., 89(14), 180
Reynolds, W. E., 133(129), 184
Reynolds, W. J., 112(118), 131(118), 171(118), 184
Richter, V. J., 154(182), 186
Riley, R. L., 10, 76
Rindfleisch, T. C., 112(120), 133(120), 135(120), 161(120), 184
Risinger, B. H., 38(105), 51(105), 79
Ritchie, J. M., 86(9), 180
Roach, M. K., 89(26), 101(82), 105, 106, 181, 183
Roberts, C. B., 38(93), 54(93), 78, 154(179), 186
Robertson, D. N., 57(134), 80
Robinson, A. B., 153, 154(172), 154(185), 158(185), 159(185), 160(185), 161(170), 161(185), 161(228), 161(229), 165(185), 170, 170(185), 171(228), 186, 188
Roboz, J., 81
Rockerbie, R. A., 101(73), 182
Roggin, G., 89(22), 181
Rollo, I. M., 152(168), 186
Rooth, G., 154(183), 186
Roszel, N. O., 134(142), 138(142), 139(142), 171(142), 185
Roughton, F. J. W., 10(34), 10(35), 76
Rubin, E., 85(2), 89, 89(34), 89(35), 89(36), 89(37), 89(38), 89(39),

89(40), 89(41), 89(42), 89(45), 89(46), 89(48), 90, 91, 92, 180, 181, 182
Runquist, O., 82
Russ, C., 20(58), 28(58), 33, 35, 77
Ruthven, C. R. J., 161(226), 188
Rutledge, C. O., 38(90), 47(90), 51, 51(90), 53, 54, 78
Ruveda, M. A., 3(17), 72, 73, 76
Rydberg, W., 104, 183
Ryhage, R., 82, 112(113), 128(113), 184

S

Salanitre, E., 38(92), 38(126), 53(92), 78, 79
Sandler, M., 161(226), 188
Sasame, H. A., 89(30), 89(31), 181
Sass, S., 3(18), 71, 76
Saunders, R. A., 82
Savory, J., 134(142), 138, 139, 171(142), 185
Schenkman, J. B., 89(12), 180
Scherer, O., 38(121), 48(121), 79
Schloegel, E. L., 36(73), 38(73), 77, 134, 134(133), 136(133), 140(133), 161(133), 171(133), 184
Scholz, R., 170(235), 188
Schultz, T. H., 171(241), 189
Schwartz, S., 89(17), 180
Scipioni, A., 38(118), 48(118), 79
Scott, R. P. W., 75, 79
Sedgwick, P. R., 101, 101(60), 182
Seifen, E., 38(90), 47(90), 51(90), 54(90), 78
Seixas, F., 96, 182
Sendroy, J. (Tr.), 11, 76
Shah, M. S., 3(21), 76
Shanley, B. C., 89(16), 89(20), 180
Shargel, L., 59, 80
Sheid, B., 31(71), 77
Shepard, R. H., 10(31), 76
Sher, R., 38(89), 48(89), 49(89), 50(89), 78
Shore, P. A., 83
Shunbo, F., 161(222), 161(223), 165(222), 165(223), 168(222), 168(223), 169(223), 170(223), 188
Shurlock, B. C., 81
Sie, S. T., 161(221), 162, 188
Siker, E. S., 38(115), 51(115), 79
Silverstein, R. M., 82
Simpson, G. S., 105(80), 112(106), 120(80), 120(106), 131(80), 131(106), 171(80), 171(106), 183
Sine, H. E., 112(114), 126, 128, 129, 171(114), 184
Sjovall, J., 38(104), 51(104), 79
Skrabvanek, P., 101(71), 182
Smalldon, K. W., 112(118), 112(119), 131(118), 133, 171(118), 171(119), 184
Smith, H. W., 120, 155(199), 184, 187
Smith, J. R., 10(28), 76
Smith, R. N., 3, 4(22), 76
Smith, R. V., 38(95), 78
Solomonraj, G., 134(149), 143(149), 185
Solon, J., 112(112), 125, 161(112), 184
Spector, N. H., 154, 187
Spiteller, G., 82
Spitz, H. D., 3(14), 63(14), 64, 65, 75
Stanovick, R. P., 69, 80
Steed, E., 133(129), 184
Stein, S., 89(21), 180

Steinberg, M., 101(69), 103, 171(69), 182
Stellate, R. L., 20(55), 29, 31, 32, 33, 34, 77, 134(134), 134(135), 137, 138, 139, 161(134), 161(135), 171(134), 171(135), 185
Stembridge, V. A., 20(64), 36(64), 77
Stewart, C. P., 182
Stewart, R. D., 38(93), 54, 78, 154(177), 154(179), 186
Stewart, W. W., 184
Stober, H. C., 20(43), 76
Stock, K., 91, 181
Stokke, O., 161(225), 188
Stolman, A., 77, 182
Stoner, E., 171(251), 177, 179, 189
Storrs, E. E., 111(89), 183
Stromme, J. H., 150(157), 150(159), 185
Sturner, W. Q., 112(105), 183
Suckling, C. W., 38(116), 48(116), 50(116), 52(116), 79
Sudlow, M. F., 59(137), 80
Sulway, M. J., 154(184), 157, 158(211), 158(212), 171(184), 171(211), 171(212), 186, 187
Summers, F. W., 38(75), 42, 78
Summons, R. E., 112(120), 133(120), 135(120), 161(120), 184
Sundby, P., 85(5), 180
Sunderman, F. W., Jr., 134(142), 138(142), 139(142), 171(142), 185
Sunshine, I., 128, 184
Svec, H. J., 20(46), 76
Svoboda, D. J., 89(13), 180
Swank, J. D., 38(93), 54(93), 78, 154(179), 186
Swinehart, J., 3, 76
Swinnerton, J. W., 2(7), 75
Szulczewski, D. H., 4, 76
Szymanski, H. A., 78

T

Takabatake, E., 89(14), 180
Taylor, B., 20(58), 28(58), 33(58), 35, 77
Tenney, H. M., 115(121), 184
Tephly, T. R., 89(27), 181
Teranishi, R., 152(170), 154(185), 158, 158(213), 158(214), 159(215), 159, 159(185), 160, 161(170), 161(185), 161(228), 161(229), 165, 170, 170(170), 171(228), 186, 187, 188
Terfloth, H. P., 134(145), 140, 185
Tewari, S., 96, 182
Theis, R. C., 162(231), 188
Theye, R. A., 38(91), 47(91), 51(91), 53, 78
Thomas, D. J., 20(62), 77
Thompson, B., 20(63), 77
Thompson, W. E., 154(173), 155, 155(173), 156, 161(173), 186
Thornton, J. A., 38(111), 46(111), 56(111), 79
Thurman, R. G., 170(235), 188
Tiengo, M., 38(112), 79
Tinelli, F., 89(27), 181
Tishbee, A., 154(186), 159(186), 161(186), 161(222), 161(223), 165(222), 165(223), 168(186), 168(222), 168(223), 169(223), 170(223), 171(186), 186, 188
Tobon, F., 89(22), 181
Tonzetich, J., 154(182), 186
Trenholm, H. L., 134(149), 143(149), 185

Triolo, G., 38(118), 48(118), 79
Trotter, E., 107, 154(184), 157, 158(212), 171(184), 171(212), 186, 187
Trotter, M. D., 154(184), 157, 158(212), 171(184), 171(212), 186, 187
Truitt, E. B., 107, 134(131), 140, 141, 171(83), 171(131), 183, 184
Tscguddy, D. P., 89(19), 180
Tucker, R. B., 133(129), 184
Turner, M., 153(170), 161(170), 170(170), 186

U

Ueda, N., 155(197), 187

V

Vanden Heuvel, W. J. A., 159, 188
Van Dyke, R. A., 38(106), 47, 79
Van Haaften, A. B., 161(221), 162, 188
Van Note, W., 154(188), 155(188), 186
Van Slyke, D. D., 11, 29, 35(72), 76, 77
Van Stekelenburg, G. J., 57, 80
Viguau, J. A., 112(103), 183
Viney, D. J., 38(116), 48(116), 50(116), 52(116), 79
Virtanen, A. I., 175(252), 189
Vitenberg, A. G., 134(153), 148, 150, 185
Von Wartburg, J. P., 89(28), 181

W

Wada, O., 89(15), 180
Wagner, H. J., 112(102), 183
Walker, A. R., 103(77), 182
Walker, G. W., 105(80), 112(106), 120(80), 120(106), 131(80), 131(106), 171(80), 171(106), 183
Walker, J. Q., 101(69), 103(69), 171(69), 182
Walker, S. R., 59(137), 80
Wallace, J. E., 128, 134(138), 134(139), 137, 140, 141, 161(139), 171(139), 184, 185
Waller, G. R., 81
Walsh, M. J., 91(50), 92, 92(52), 170(237), 170(238), 182, 188, 189
Watkins, J., 112(112), 125(112), 161(112), 184
Watkins, W. D., 89(27), 181
Watson, J. T., 82
Way, E. L., 83
Wehrli, H., 154(173), 155(173), 161(173), 186
Weigand, W., 38(121), 48(121), 79
Weinberger, J., 3(14), 63(14), 64, 65, 75
Weiner, M., 85(4), 180
Wells, J., 150(166), 152, 186
Welton, W. M., 3(20), 76
Wesselman, H. J., 112(88), 115, 116, 183
West, W. B., 83
Westerfeld, W. W., 87(10), 180
Westermann, E., 91, 181
Whelpton, D., 38(111), 46(111), 51(111), 79
Whitnah, C. H., 171(250), 189
Wiberg, G. S., 134(149), 143(149), 185
Wibowo, A. E., 38(98), 57(98), 60(98), 78
Williams, A. E., 82
Williams, D. H., 38(84), 48(84), 78, 81, 82
Williams, F. M., 59(136), 80
Williams, P. M., 155(203), 187

Williams, R., 154(173), 155(173), 161(173), 186
Williamson, J. R., 170(235), 188
Willis, R., 20(46), 76
Wilson, C. J., 89(17), 180
Wilson, C. O., 83
Wilson, R. H., 10(26), 11, 12, 13, 14, 20, 20(26), 20(49), 20(51), 20(65), 21(26), 21(49), 21(51), 21(65), 22, 23, 23(49), 23(51), 23(65), 24, 76, 77
Winek, C. L., 128, 184
Witten, T. A., 161(227), 168, 170, 170(233), 173, 188
Wolf, G. L., 38(92), 38(126), 53(92), 78, 79
Wolfson, B., 38(115), 51(115), 79
Wood, D. G. L., 38(119), 51(119), 54(119), 79, 171(248), 189
Woods, E. F., 38(97), 57(97), 78
Worm, K., 38(114), 48(114), 79
Wortley, D. J., 38(111), 46(111), 51(111), 79
Wright, B. M., 155(196), 187

Y

Yamamoto, K., 155(197), 187
Yamanaka, Y., 92(52), 170(238), 181, 189
Yarger, K., 86(7), 161(7), 163(7), 171(7), 180
Yee, H. Y., 20(52), 28, 77
Yee, L. L., 112(98), 117(98), 118(98), 119(98), 161(98), 171(98), 183
Yokota, T., 38(123), 51(123), 79
Yuzo, Y., 89(15), 180

Z

Zack, J. F., Jr., 171(247), 177(247), 189
Zail, S. S., 89(16), 89(20), 180
Zapp, J., 38(122), 51(122), 79
Zauder, H. L., 38(77), 38(79), 42, 78
Zerba, E. N., 3(17), 72, 73, 76
Ziegler, D. M., 89(23), 181
Zieve, L., 89(17), 180
Zieve, P., 89(22), 181
Zlatkis, A., 80, 154(186), 159, 159(219), 160, 161, 161(186), 161(219), 161(222), 161(223), 165, 168, 169, 170, 171, 186, 188

SUBJECT INDEX

A

Acetaldehyde, 86, 87, 92, 93, 96, 105, 106, 112, 119, 120, 121, 124, 130, 136, 140, 145, 149, 156, 157, 161, 170, 175, 177, 178
  dehydrogenase, 87, 89, 92, 93
  recovery of, 106
Acetic acid, 87, 116, 175, 176
  detection limits of, 176
Acetone, 3, 57, 64, 65, 74, 75, 102, 106, 112, 115, 117, 119, 120, 121, 124, 126, 128, 129, 136, 137, 140, 145, 149, 152, 156, 157, 158, 161, 162, 169, 175, 177, 178, 179
  recovery of, 121
Acetonitrile, 119, 153, 156, 177
Acetophenone, 72, 174
Acetyl coenzyme (CoA), 87
Acetylene, 4, 5, 6, 7, 9, 10, 12, 14, 18, 24, 26, 36, 37
  rate of disappearance in alveolar air, function of, 9
Acetylene dichloride, 119
Acetylfuran, 169
Aconitic acid, 173
Acrolein, 119
Acrylonitrile, 66, 67, 119
Activated charcoal, 156
Activity coefficient, 149, 150
Adamsite, see Diphenylaminechloroarsine
Adenosine triphosphate (ATP), 86, 87
Adipic acid, 170, 173
Adsorbent Chromosorb 101 precolumn, 171
Adsorbent Tenax GC trap, 159, 160, 166
Alcohol(s), 117, 141-143
  dehydrogenase (ethyl alcohol), see Ethyl alcohol
  nitrite esters of, using nitrous acid as reagent, 117, 141-143
Alkylfuran(s), 168
Allo-tetrahydrocorticosterone, 165
Allo-tetrahydrocortisol, 165
Allyl alcohol, 119
Allyl ether, 119
Allyl isothiocyanate, 168,169
Alveolar air, 8, 9, 11, 14, 86, 154
Alveolar-capillary carbon monoxide gradient, 8
Alveolar-capillary membrane, 8, 9
Alveolar volume, 8
Aminopyrene demethylase, 91
Amitriptyline (Elavil), 127
Ammonia, 4
Amobarbital, 108, 109, 110, 127
  recovery of, 109

Amylacetate, 136
Amyl alcohol, 119
tert-Amyl alcohol, 109, 110
5-Androsten-3β,16α-diol-17-one, 166
  TMSi derivative, 166
5-Androsten-3β,17β-diol-16-one, 166
  TMSi derivative, 166
5-Androsten-3β,16α,17β-triol, 166
  TMSi derivative, 166
Androsterone, 165
Anesthetics, 38-62
  bioavailability from inhalation, 38
  Ostwald partition coefficients of common, 45
  physical properties of common, 45
  typical blood concentrations of common, 45
Aniline hydroxylase, 90, 91
Antabuse, see Disulfiram
Antabuse reaction, 152
Antifoam[60], 22
Apparent diffusion capacity of lung, see Diffusing or diffusion capacity (lung)
Apparent volume of distribution, 38
Argon, 11, 12, 13, 15, 21, 36, 37, 48, 51, 72, 120, 135
  comparison of data for, by GC and Van Slyke methods, 11
  metastable, 37
Arrhenius plot, 133
Ascorbic acid, 164
Atabrine, see Quinacrine
Automatic capsule sampler, 130, 131
Automatic liquid sampler, 125, 126
Automatic sample injector, 125, 126, 144-148
Automatic sampling valve, 5

B

Barbital (Veronal), 127
Barium hydroxide, 106, 140
Benzaldehyde, 169, 174, 179
Benzene, 57, 70, 71, 117, 119, 136, 162, 163, 178
Benzpyrene hydroxylase, 90
Benzyl alcohol, 131
Benzyl bromide, 72, 73
Benzyl cyanide, 71
Bis-(trimethylsilyl)trifluoroacetamide, 167
Blood alcohol analyzer, 125
  description of, 125
Blood analysis (volatiles), 171-180
Blood gas analysis, 21-38
  carbon dioxide concentration in blood, 24
    equation for, 24
  GC instruments for, 21-38
    description of methods used, 21-38
    glass cuvette of, 23
    reaction chamber of, 25
    schematics of, 22, 25
  oxygen concentration in blood, 24
    equation for, 24
Body weight, 86, 169
Borax, 35
Borosilicate glass powder, 163
Breath analysis, 154-160
Breath trap (volatiles), 158
  schematic of, 158
Bromoacetone, 74, 75
Bromobenzene, 66, 67
Bromobenzylcyanide, see α-Bromobenzylnitrile
α-Bromobenzylnitrile, 71, 72
  detection limits of, 72
1-Bromobutane, 66, 67

2-Bromobutane, 66, 67
2-Bromo-2-chloro-1,1-difluoroethane, 52
2-Bromo-2-chloro-1,1,1-trifluoroethane, see Halothane
Bromodichlorofluoromethane, 52
2-Bromo-2,2-dichloro-1,1,1-trifluoroethane, 52
Bromoform, 66, 67, 68
2-Bromohexafluorobutene-2, 50
2-Bromo-1,1,1,4,4,4-hexafluorobutene-2 (trans), 52
1-Bromopentane, 66, 67
2-Bromopentane, 66, 67
1-Bromopropane, 66, 67, 68
2-Bromopropane, 66, 67, 68
3-Bromopropane, 66, 67
3-Bromo-1,1,2,2-tetrafluoropropane (Halopropane), 38, 45, 56, 57, 58-59
  comparison of solvent extraction/direct injection GC methods for, 58, 59
  physical properties of, 45
  recovery from blood (direct injection), 57, 58-59
Bromotoluene (o- and p-isomers), 72, 73
2-Bromo-1,1,1-trifluoroethane, 52
Bunsen solubility coefficient, 9, 29, 30, 31
Butabarbital (Butisol), 108, 109, 110, 127
  recovery of, 109
n-Butane, 19
2,3-Butanedione, 169
n-Butanol, 44, 102, 103, 104, 107, 110, 111, 115, 119, 136, 142, 143, 145, 149, 169, 179
sec-Butanol, 119, 142
tert-Butanol, 106, 119, 142, 144, 145, 146, 148, 149
  vapor pressure of, function of temperature, 144
2-Butanone, 119, 168, 169, 179
n-Butyl acetate, 119
n-Butyl amine, 119
n-Butyl chloride, 152
n-Butyl ether, 119
n-Butyl nitrite, 142, 143
sec-Butyl nitrite, 142
tert-Butyl nitrite, 142
1-Butylpyrrole, 169
n-Butyric acid, 140, 164
n-Butyronitrile, 119

C

Calcium carbimide, 89
Calcium chloride, 153
Calcium cyanamide, 89
Calcium hydride, 117
Calcium sulfate, 32, 102, 103, 111
Calibration factor, 148
  determination of, using peak height measurement, 148
Camphor, 131
Capillary blood flow, see Lung capillary blood flow
Caprylic alcohol, 35
Carbon dioxide, 1, 2, 3, 4, 5, 6, 7, 10, 12, 13, 14, 15, 16, 17, 18, 20, 23, 24, 25, 27, 28, 29, 33, 34, 35, 36, 37, 39, 40, 48, 86, 87, 88
  adsorption by charcoal of, 3, 4
  adsorption by molecular sieve of, 12
  calibration curve of, 13
  comparison of data obtained by GC and Van Slyke methods, 11, 28
Carbon disulfide, 66, 68, 119, 150, 151, 152
Carbon monoxide, 2, 3, 4, 5, 6, 7, 8, 9, 10, 11, 12, 14, 17, 18, 19, 36, 37

[Carbon monoxide]
calibration curve of, 12
capacity, 36
complex with hemoglobin (carboxyhemoglobin), 8, 36
equation for determination of, 36
rate of combination, oxygen partial pressure-dependency, 8
Carbon tetrachloride, 3, 41, 66, 68, 69, 70, 71, 119, 136, 175
Cardiac output, 9, 10, 11, 53, 54
equivalent to lung capacity blood flow, 9
Carvone, 168, 169, 174
Charcoal adsorption column, 3, 4
adsorptivity of carbon dioxide on, 3, 4
reactivity of, 4
Chloral hydrate, 95, 108, 109, 127, 136
effect of alkali on, 136
recovery of, 109
Chloramphenicol, 94
recovery of, 109
Chlordiazepoxide (Librium), 108, 109, 110, 127
o-Chloroacetone, 74, 75
o-Chloroacetophenone, 71, 72, 73
detection limits of, 72
o-Chlorobenzaldehyde, 72
o-Chlorobenzalmalonitrile, 71, 72, 73
detection limits of, 72
Chlorobenzene, 66, 67, 68, 70, 71, 119
2-Chloroethanol (ethylene chlorohydrin), 64, 65, 119
detection limits of, 64
Chloroform, 1, 3, 4, 38, 39, 41, 42, 43, 45, 48, 52, 59, 66, 68, 70, 71, 107, 119, 136, 145, 169, 175, 178, 179
^{14}C-labeled, 48
physical properties of, 45
2-Chloro-1,1,1,4,4,4-hexafluorobutene-2 (cis), 52
2-Chloro-1,1,1,4,4,4-hexafluorobutene-2 (trans), 52
Chloromycetin, 94
1-Chloro-2-nitropropane (Lanstan), 69
detection limits of, 69
recovery of, 69
Chloropicrin, 66, 67
2-Chloro-1,1,1-trifluoroethane, 50, 52
Chlorpromazine (Thorazine), 127
Chlorpropamide (Diabinese), 94
Cholesterol, 87, 90, 165
Cholesteryl butyrate, 165, 166
Citric acid, 87, 89, 167, 173
cycle, 87
Column packing (solid substrate), types of
Anakrom-series, 1, 41
Carbosieve-B, 87, 88
Celite-series, 39, 50, 144, 120, 137, 144
Charcoal, 10, 13, 14, 24
Chromaton NAW, 20, 47
Chromosorb (porous polymer), 18, 19, 141, 143, 157
Chromosorb-series (diatomite), 1, 11, 12, 14, 24, 25, 26, 41, 42, 43, 46, 47, 48, 49, 50, 55, 63, 69, 74, 102, 103, 107, 111, 115, 118, 124, 137, 138, 153, 163, 176, 178
Diatoport-series, 66, 104, 120
Embacel, 175
Firebrick, 4, 10, 28, 39, 40, 41, 44, 47, 56, 57, 64, 102, 110,

[Column packing (solid substrate), types of]
[Firebrick] 115, 135, 138, 139
Fluoropak 60-F, see Haloport-60F
Gas Chrom-series, 71, 164, 165
Glass wool, 43
Haloport-60F, 143, 176
Kieselguhr, 117
Molecular sieve-series, 5, 6, 7, 10, 11, 12, 13, 18, 19, 24, 28, 31, 35, 36, 46, 47, 156
Porapak-series, 18, 20, 47, 59, 69, 70, 71, 74, 75, 107, 109, 110, 121, 122, 123, 125, 126, 129, 130, 132, 133, 142, 144, 145, 155, 156, 177
Polypak-2, 106
Silica gel, 3, 4, 5, 6, 7, 14, 25, 26, 28, 31, 41, 47, 54, 115, 117
adsorptivity of, 3
Spherosil-XOB-15, 69
Sterchamol, see Silica gel
Styrene-divinyl copolymer, 64
Teflon-series, 47, 63, 118, 124, 137, 139
Tide, 43, 47, 49, 55
Varaport-30, 72, 152
XAD-resin, 20, 21
Component concentration in column liquid stationary phase, 148-150
for dilute solutions, equation for, 149
initial, equation for, 149
Copper sulfate, 137, 138, 175
Cortol, 165
β-Cortol, 165
Cortolone, 165
β-Cortolone, 165
Creatinine, 172
Cresol(s), 162, 163
m-Cresol, 162, 163
p-Cresol, 162, 163, 169
Cross-section ionization detector, see Detector, cross-section ionization
Cyclohexane, 41, 61, 62, 119, 177
Cyclohexanone, 168, 174
Cyclopropane, 1, 2, 3, 4, 5, 24, 26, 37, 38, 40, 42, 43, 45, 59, 62
physical properties of, 45

D

Data processor hardware, 125
1-Decanol, 174
Dehydroepiandrosterone, 165
Dehydroisoandrosterone, 107, 109, 110
Detector
cross-section ionization, 1, 2
electron capture, 51, 54, 55, 59, 69, 71, 72, 168
flame ionization, 38, 39, 42, 43, 44, 48, 49, 50, 64, 69, 71, 72, 74, 103, 106, 107, 115, 117, 118, 120, 121, 122, 123, 124, 125, 126, 130, 132, 137, 139, 141, 143, 146, 152, 153, 155, 156, 157, 158, 161, 162, 163, 164, 171, 175, 177, 178
flame photometric, 168
for sulfur detection, 168
microcoulometric, 69
radium-226 beta ionization, 134
thermal conductivity, 1, 2, 3, 5, 11, 15, 19, 20, 21, 25, 31, 38, 39, 40, 42, 49, 64, 66, 71, 72, 74, 82, 102, 103, 110, 111, 114, 115, 117, 137, 139, 142, 177
Diacetyl, 156
Diazepam (Valium), 95, 108, 109, 110, 127
recovery of, 109
1,1-Dibromacetone, 74, 75

1,2-Dibromo-3-chloropropane, 69
2,2-Dibromo-2-chloro-1,1,1-trifluoroethane, 52
1,10-Dibromodecane, 71
1,3-Dibromopropane, 66, 67
2,2-Dibromo-1,1,1-trifluoroethane, 52
1,1-Dichloroacetone, 74, 75
o-Dichlorobenzene, 66, 67
p-Dichlorobenzene, 66, 67, 68
1,2-Dichloro-1,1-difluoroethane, 52
2,2-Dichloro-1,1-difluoroethylmethyl ether (methoxyflurane, perthane), 38, 43, 45, 48, 56, 57, 59
  ^{14}C-labeled, 48
  physical properties of, 45
Dichlorodifluoromethane (Freon 12), 56, 59, 60, 62, 63, 74, 75
  detection limits of, 60
  partition coefficient of, in, 60
Dichloroethane, 70, 71
Dichloroethylene, 70, 71
β,β-Dichloroethyl ether, 66, 67, 68
Dichlorohexafluorobutene-2 (cis/trans), 46, 48, 49, 50, 51, 52
  formation of, 48, 49
2,3-Dichloro-1,1,1,4,4,4-hexafluorobutene-2 (cis), 48, 49, 50, 51, 52
2,3-Dichloro-1,1,1,4,4,4-hexafluorobutene-2 (trans), 48, 49, 50, 51, 52
1,2-Dichloro-1,1,2,2-tetrafluoroethane (Freon 114), 56, 59, 60, 62
  detection limits of, 60
  partition coefficient of, in, 60
2,2-Dichloro-1,1,1-trifluoroethane, 52
Diethylamine, 119, 151, 152
Diethyldithiocarbamic acid, 150, 151, 152
  methyl ester of, 150, 151, 152
N,N-Diethyldithiocarbamoyl-1-thio-β-D-glucopyranosiduronic acid, 150, 151
Diethyl ether, 1, 2, 3, 4, 6, 38, 39, 40, 41, 42, 43, 45, 48, 59, 70, 71, 102, 112, 114, 119, 120, 121, 136, 145, 149, 152, 166, 175, 176, 177, 179
  ^{14}C-labeled, 48
  physical properties of, 45
  recovery of, 177
Diethyl sulfide, 156
Diffusing or diffusion capacity (lung), 5, 8, 9, 10, 12, 14, 17
  calculation of apparent, 8
    definition of, 8
    equation for, 8
  of carbon monoxide, defined as, 8
  definition of, 8, 14
  determination of, 5, 8, 17
  simultaneous analysis of, with lung capillary blood flow, 14
  two-stage transfer process, 8
    chemical association, 8
    passive membrane diffusion, 8
3,4-Dihydroxyphenylacetaldehyde, 93
Dihydroxyphenylpropionic acid, 173
Dimethylchlorosilane, 74
Dimethyldisulfide, 169
2,3-Dimethylfuran, 169
2,4-Dimethylfuran, 169
2,3-Dimethylpyrazine, 169
2,5-Dimethylpyrazine, 169
2,6-Dimethylpyrazine, 169
Dimethylpyrrole, 169
Dimethylsulfide, 156, 168
Dimethylsulfone, 168
Dioxane, 119, 120

Diphenylamine, 72, 73
Diphenylaminechloroarsine (Adamsite), 72
Diphenylhydantoin (Dilantin), 94, 108, 109, 110, 128
recovery of, 109
Direct GC method for ethyl alcohol analysis, 101, 112-134
categories of
dilution-injection, 114, 115, 116, 117, 120, 125, 126, 128, 129, 130, 134
direct (as is), 114, 115, 118, 120, 121, 122, 124, 130, 131, 132
precipitation (protein)-direct, 114, 123
Disodium phosphate, 170
Distillation-GC method for ethyl alcohol analysis, 101, 105, 110-112, 114, 115, 116, 120
distillation apparatus for, 112
Distribution coefficient, see Partition coefficient
Disulfiram, 89, 94, 95, 150, 151, 152
metabolic pathways of, 151
metabolic products of, 150, 151
structure, 89, 151
Divinyl ether, 38, 42, 59
1-Dodecanol, 174
Dopamine, 92, 93
structure, 93
Double-helix trap, 159, 171
schematic of, 159

E

Eicosane, 174
Electron capture detector, see Detector, electron capture
Electronic integrator, 125, 126
Electropneumatic dosing system, 144, 146-148
description of, 146-148
schematic of, 147
Emphysema, 12, 16
abnormal GC pattern for, 12, 16
normal GC pattern for, 12, 16
Ethane, 19, 36, 37
Ethanol, see Ethyl alcohol
Ethchlorvynol (Placidyl), 108, 109, 110, 127
recovery of, 109
Ethyl acetate, 118, 119, 136
Ethyl alcohol, 3, 41, 85, 86-153, 154, 155, 156, 157, 161, 162, 168, 169, 170, 171, 172, 175, 177, 178, 179
binding of, to microsomal P-450, 90
blood concentration of, 97, 100, 101, 124
equations for, 97, 100, 124
plot of time evolution of, following constant rate of alcohol ingestion, 100
single alcohol dose ingestion, 98
time-dependency of, following single alcohol dose ingestion, 97, 98, 99
blood level of, dependent on
amount ingested, 86
body weight, 86
rate of absorption, 86
rate of distribution, 86
rate of metabolism, 86
total body weight, 86
concentration in body fluids, relative to blood, 86
decay profiles for blood and urine, 134, 135
dehydrogenase, 86, 87, 104, 105, 130, 131, 141
detection limits for, 134, 139, 141

[Ethyl alcohol]
deuterated, for MID detection, 134
drug interactions with, 91, 94-95
effect of bacteria on, in post mortem samples, 140-141
effect of sodium chloride on vapor pressure of, 138
effect on hepatic drug-metabolizing enzyme activity, 89, 90
effect on pentobarbital metabolism, 92
effect on smooth endoplasmic reticulum, 89
elimination time, 99
definition of, 99
equation for, 99
formation of nitrite derivative of, 141-143
gas chromatographic methods of analysis for, via, 101-150
direct injection, 101, 112-134
distillation-injection, 101, 105, 110-112, 115, 116, 120
extraction-injection, 101-110
headspace, 101, 134-150
inhibition of hepatic drug-metabolizing enzymes by, 90, 91
metabolism of, 87, 89, 96-101, 105, 141
altered by, 89
overall pharmacokinetic mechanism for, 96-101
conclusions drawn from, 101
non-GC method of analysis for, 155
chemical, 155
conductance, 155

[Ethyl alcohol]
[non-GC method of analysis for]
dehydrogenation, 155
electrochemical, 155
heat of oxidation, 155
infrared absorption, 155
quantitative determination of, 124, 148
equations for, 124, 148
rate of adsorption of ingested, influenced by, 86
concentration, 86
food in stomach, 86
rate of oxidation, 86
linear function of time, 86
nearly proportional to body and liver weight, 86
sobriety time, 99, 101
definition of, 99
equation for, 99
specific instruments for analysis of, 142, 155
Alco-Analyzer, 142, 155
Intoximeter, 155
study of oxidation of, 133
study of stability of, influenced by, 131, 132, 133
alcohol concentration, 131, 132, 133
fluoride concentration, 131, 132, 133
temperature of storage, 131, 132, 133
time of storage, 131, 132, 133
type of container, 131, 132, 133
time evolution of, 97, 98, 100
time for maximum blood concentration of, 99
equation for, 99
vapor pressure for, function of temperature, 144
Ethyl bromide, 67
2-Ethylbutyraldehyde, 119
Ethyl chloride, 38, 41

Ethylene, 24, 26, 37, 38, 42, 117
Ethylene chlorohydrin, see 2-Chloroethanol
Ethylene dibromide, 66, 67, 69
Ethylene dichloride, 66, 68, 69, 119, 136
Ethylene glycol, 64, 65
  detection limits of, 64
Ethylene oxide, 63, 64, 65, 66
  detection limits of, 64
Ethylene trichloride, 119
Ethyl iodide, 119
Ethylmorphine demethylase, 91
Ethyl nitrite, 141, 142, 143
Eticholanolone, 165
Eucalyptol, 131
Extraction-GC method for ethyl alcohol analysis, 101-110

F

Fatty acids, 87, 164
Flagyl, see Metronidazole
Flame ionization detector, see Detector, flame ionization
Flame photometric detector, see Detector, flame photometric
Fluoromar (trifluoroethyl vinyl ether), 38, 42, 43, 44, 45, 46, 47, 59, 111
  physical properties of, 45
  recovery of, 47
Fluothane, see Halothane
Flurazepam (Dalmane), 108, 109, 110, 127
  recovery of, 109
Formaldehyde, 87, 88, 111, 119, 136, 178
Formic acid, 87, 88, 119, 176
Four-way sample valve, 6, 7, 18, 19, 25, 38, 160, 161
Fraction of recovery, 62
Freon 11, see Trichlorofluoromethane
Freon 12, see Dichlorodifluoromethane
Freon 113, see 1,1,2-Trichloro-1,2,2-trifluoroethane
Freon 114, see 1,2-Dichloro-1,1,2,2-tetrafluoroethane
Fumaric acid, 167
Furan, 156
2,5-Furandicarboxylic acid, 170, 173
Furfural, 169

G

Gas phase volume (column), 149
Glucose, 157
β-Glucuronidase, 171
Glusulase, 165, 166, 171
Glutethimide (Doriden), 107, 108, 109, 110, 127
  recovery of, 109
Glycerol, 164
Glycine, 130

H

Halopropane, see 3-Bromo-1,1,2,2-tetrafluoropropane
Halothane, 1, 2, 3, 4, 6, 20, 38, 39, 40, 42, 43, 44, 45, 46, 47, 48, 49, 50, 51, 52, 53, 54, 55, 59, 62, 111, 175
  ^{14}C-labeled, 46, 47
  concentration in blood, 53, 55
    arterial, 55
    capillary, 55
    dog (time-course study), 53, 54
    equation for, 53
    venous, 55
  concentration in expired gas (dog), time-course, 54
  concentration in inspired gas (dog), time-course, 54
  detection limits of, 39
  impurities in, 46, 48, 50, 51, 52

[Halothane]
physical properties of, 45
recovery of, 47
tension in blood, 2, 48
Headspace-GC method for ethyl alcohol analysis, 101, 134-150
Helium, 3, 4, 10, 11, 12, 13, 14, 15, 19, 21, 23, 24, 25, 27, 31, 32, 35, 36, 37, 38, 40, 41, 42, 48, 57, 64, 67, 71, 74, 88, 102, 103, 104, 106, 107, 110, 114, 115, 120, 123, 133, 137, 138, 139, 158, 160, 161, 177
calibration curve for, 13
equilibrium chamber, 27
Hematocrit, 33, 59
Hemoglobin, 8, 21, 23, 33, 35, 36
carbon monoxide complex of (carboxyhemoglobin), 8, 36
equation for the determination of, 36
rate of combination, oxygen partial pressure-dependency, 8
Heparin, 143
Hepatic catalase, 87, 88
Hepatic microsomes, 89, 90
Heptadecane, 174
n-Heptane, 39, 119, 136
n-Heptanol, 117
2-Heptanone, 169
4-Heptanone, 168, 169, 179
Hexachlorobutadiene, 66, 67
n-Hexacosane, 166
Hexamethyldisilazane, 42, 103
n-Hexane, 70, 71, 119, 136
n-Hexanol, 117
3-Hexanone, 169
Human metabolic profiles, 86, 161, 163, 164, 165, 166, 168, 170, 171
Hydrogen, 4, 5, 20, 36, 37, 39, 40, 114, 117, 163
Hydrogen cyanide, 66
Hydrogen peroxide, 28
Hydrogen sulfide, 36, 150
2-Hydroxybutyrate, 176
3-Hydroxybutyrate, 157
α-Hydroxybutyric acid, 170, 172
β-Hydroxybutyric acid, 167, 170, 172
1-β-Hydroxyethyl-2-methoxy-5-nitroimidazole, 152, 153
TMSi derivative, 153
1-β-Hydroxyethyl-2-methyl-5-nitroimidazole, see Metronidazole
5-Hydroxyindoleacetic acid (5-HIAA), 96
β-Hydroxyisovaleric acid, 172
5-Hydroxy-2-methylfuroic acid, 173
p-Hydroxyphenylacetic acid, 170, 173
5-Hydroxytryptamine, 96

I

Imipramine (Tofranil), 127
Infrared spectrophotometer, 10, 54
Injection port cartridge, 121
for direct on-column injection, 121
schematic of, 121
Inositol, 164
Iodine pentoxide, 3
Ionization potential, 37, 38, 74, 166
of various gases and organic materials, 37
Isoamyl alcohol, 57, 119
Isobutyl alcohol, 119, 124, 136, 142, 143, 178
Isobutyl nitrite, 142, 143
Isobutyric acid, 140
Isocitric acid, 167, 173
Isoniazid, 94
Isopentane, 19
Isoprene, 156, 157

Isopropanol, 102, 107, 108, 109, 110, 112, 119, 120, 121, 124, 126, 128, 129, 136, 137, 140, 142, 143, 145, 157, 158, 177, 178
Isopropyl ether, 119
Isopropyl nitrite, 142, 143
Isovaleraldehyde, 156

J

Jet separator (GC-MS interface), 153, 166
  two-stage, 153

K

11-Ketoandrosterone, 165
11-Ketoetiocholanolone, 165
α-Ketoglutaric acid, 167
Krebs cycle acids, 165, 167
  MO-TMSi derivatives, 167
  TMSi derivatives, 167
Krypton, 21

L

Lactic acid, 22, 119, 138, 157, 167
Lanstan, see 1-Chloro-2-nitropropane
Legal definition of intoxication, 97, 98, 100
Leucine, 87
Liquid stationary phase, types of
  Amine 220, 47
  Apiezon L, 55, 153
  Arochlor 1254, 50, 52
  Benzylether, 39
  Bis-[2-(2-methoxyethoxy) ethyl] ether, 26
  1,2,4-Butanetriol, 64
  Carbowax-series, 41, 47, 55, 69, 70, 71, 74, 75, 114, 115, 118, 124, 137, 139, 143, 144, 145, 149, 176, 178

[Liquid stationary phase, types of]
  Castor wax, 118
  Di-2-ethylhexylsebacate, 12, 42, 44, 47
  Diglycerol, 115
  Di-isodecylphthalate, 40, 48, 102, 134, 137, 139
  Dimethyl sulfolane, 64
  Dimethyl sulfoxide, 1, 4, 6, 39, 40
  Dinonylphthalate, 1, 4, 6, 39, 40, 50, 52, 115, 117
  Dioctylsebacate, 47, 63
  Dodecylphthalate, 42
  Dowfax-9N-15, 166
  Emulphor-O, 160
  Emulphor-ON-870, 171 174
  2,2'-(2-Ethylhexamido)diethyl-di-(2-ethylhexoate) or Flexol 8NP, 101-102, 134, 137, 139
  F-60 (siloxane polymer), 164
  FFAP, 152
  Glycerol, 110
  Hallcomid M-18, 103, 120, 124, 126, 129, 137, 138, 144, 145
    composition of, 137
  Hexadecane, 57
  Hexamethylphosphoramide, 11, 26
  Igepal, 103
  Kel-F oil, 20, 47
  Methylhexylsebacate, 42
  Narcoil, 1
  OV-1, 107, 110, 164, 165, 166
  OV-17, 72, 107, 109, 110, 153, 164
  OV-22, 167
  OV-101, 131, 164
  Polyethylene glycol-series (PEG), 50, 52, 64, 102, 115, 117, 120, 134, 137, 138, 139, 156, 163
  Polywax 4000, 128
  Propylene carbonate, 39
  QF-1, 71
  SF-96 (methyl silicone), 158, 171, 174

[Liquid stationary phase, types of]
Silicone oil-series (DC), 39, 41, 42, 46, 49, 51, 56, 162, 175
Silicone rubber (SE-30), 43, 48, 49, 66, 69, 153, 164
Squalane, 41
Terephthalic acid, 176
Tetraethylene glycol methyl ether, 41
Tricresylphosphate, 110, 115, 162
Ucon-series, 63, 115, 118
XE-60, 69
Liquid stationary phase volume (column), 149
Liver weight, 86
Lung capillary blood flow, 5, 9, 14, 18
description of, 14
determination of, 14, 18
equal to cardiac output, 9
simultaneous determination of, with diffusing capacity (lung), 14
Lung capillary volume, 10

M

Magnetic-deflection mass spectrometer, 10, 74, 128-129, 153, 156, 166
Malic acid, 167
Malonitrile, 72
Mass fragmentography, see Multiple ion detection
Menthol, 131
Meperidine (Demerol), 127
Meprobamate (Miltown), 95, 107, 108, 109, 110
recovery of, 109
Mercapturic acids, 164
Metabolic profile, 86, 161, 163, 164, 165, 166, 168
compound-types found in, 164
[Metabolic profile]
practical procedural steps for determining, including, 163, 164
analysis, 164
derivatization, 164
isolation, 163
of urinary steroids, 164, 165, 166
adult, 164, 165
infant, 165, 166
Methane, 2, 17, 19, 20, 21, 36, 37, 51, 72, 156, 157
Methanethiol, 162
Methanol, 87, 88, 102, 107, 108, 109, 110, 111, 112, 119, 120, 121, 124, 126, 128, 129, 136, 137, 140, 142, 143, 145, 156, 157, 161, 162, 177, 178, 179
metabolism of, 87, 88
recovery of, 109
Methaqualone, 108, 109, 110
recovery of, 109
Methoxyflurane, see 2,2-Dichloro-1,1-difluoroethyl methyl ether
Methyl acetate, 119
β-Methyladipic acid, 170, 173
Methylamine, 122, 123
Methyl bromide, 66, 68
3-Methyl-2-butanone, 169
Methyl chloride, 66, 68
3-Methylcyclopentanone, 169
Methylene chloride, 70, 71, 119, 152, 153
Methylene dichloride, 66, 68, 152, 153
Methylene unit, 164
for characterizing peak position, 164
Methyl ethyl ketone, 115, 121, 130, 131, 136, 137, 145, 156, 178
2-Methyl-5-ethylfuran, 169
2-Methyl-6-ethylpyrazine, 169
Methylformate, 156
Methylfuran, 156
6-Methyl-3-heptanone, 169
5-Methyl-3-hexanone, 169

Methyl iodide, 119, 152
Methyl malonic acid, 172
Methyl nitrite, 142, 143
3-Methyl-2-pentanone, 169
4-Methyl-2-pentanone, 169
4-Methyl-3-penten-2-one, 169
p-Methylpropenylbenzene, 169
Methylpyrazine, 169
1-Methylpyrrole, 169
2-Methylpyrrole, 169
Methyl stearate, 107, 110
2-Methyl-6-vinylpyrazine, 169
Methyprylon (Noludar), 107, 108, 109, 110, 127
  recovery of, 109
Metronidazole (Flagyl), 89, 94, 150, 151, 152, 153
  metabolism of, 150, 151
  structure, 89, 151
  TMSi derivative, 153
Mevalonic acid, 87
Microcoulometric detector, see Detector, microcoulometric
Molecular ion, 128
Molecular weight, 45, 53
  of common anesthetics, 45
Monoamine oxidase, 95
Morphine, 91, 92
Multiple ion detection (MID), 128-129, 133, 134
  detection limits using, 134
Myristic acid, 137
  N,N-dimethylamide, 137

N

Naloxone, 93
Naphthalene, 177, 178, 179
Natelson microgasometer, 31, 32, 37
  schematic of, 32
Neon, 5, 6, 7, 9, 10, 12, 14, 17, 18, 21
Nicotinamide adenine dinucleotide, 87, 130
Nitric oxide, 3, 4
Nitrogen, 1, 2, 3, 4, 5, 6, 7, 10, 11, 12, 13, 15, 16, 17, 18, 19, 21, 24, 25, 29, 30, 31, 34, 36, 37, 50, 59, 60, 69, 72, 117, 118, 121, 123, 124, 125, 126, 132, 138, 141, 143, 144, 152, 153, 155, 157, 158, 177, 178
  calibration curve for, 12, 13
  solubility in water, relationship with temperature, 29, 30
  wash-out curve, 5
Nitrogen dioxide, 3
  irreversible adsorption by silica gel, 3
Nitro reductase, 90
Nitrous acid, 117, 141, 142, 143
Nitrous oxide, 1, 2, 3, 4, 6, 10, 11, 13, 14, 20, 24, 26, 38, 39, 40, 42, 53
  calibration curve for, 13
2-Nonanone, 169

O

Octadecane, 174
1-Octadecanol, 174
1-Octanol, 174
2-Octanone, 174
3-Octanone, 169, 174
Octyl alcohol, 26
Olefins, 117
Oleic acid, 137
  N,N-dimethylamide, 137
Orinase, see Tolbutamide
Orthomolecular diagnosis, 154
  definition of, 154
Orthomolecular therapy, 154
  definition of, 154
Ostwald partition coefficient, 45
  of common anesthetics, 45
Ostwald solubility coefficient, 27
Oxalacetic acid, 167
Oxalic acid, 141
Oxazepam, 108, 109, 110
  recovery of, 109

Oxygen, 1, 2, 3, 4, 5, 6, 7, 9, 10, 11, 12, 13, 15, 16, 17, 18, 19, 20, 21, 23, 24, 25, 28, 29, 31, 33, 34, 35, 36, 37, 39, 40, 44, 49, 60
calibration curve for, 13
comparison of data for, by GC and Van Slyke methods, 11, 28, 35, 36
GC data for, compared with X-ray and gasometer values, 31, 33
tension, 10, 35
Oxyhemoglobin, 133

P

Paraldehyde, 3, 112, 119, 120, 124, 136, 145, 149, 175
effect of acid on, 136, 175
Partial pressure, 8, 9, 10, 27, 53, 150
relationship to peak height or area, 150
Partition coefficient, 3, 38, 48, 59, 60, 61, 62, 148, 149, 150
calculated from GC peak height, 62
expressed in concentration terms, 61
for fluorocarbons (GC method), 61, 62
equation for, 61, 62
Ostwald, 45
of common anesthetics, 45
of protein binding, 61
expressed as, 61
temperature-dependent, 150
Pentadecane, 174
n-Pentane, 19, 67, 70, 71, 119
2-Pentanone, 169
Pentazocine (Talwin), 127
Pentobarbital (Nembutal), 92, 108, 109, 110, 127
effect of ethyl alcohol on metabolism of, 92
recovery of, 109
Pentobarbital hydroxylase, 90, 91
2-n-Pentylfuran, 169
Perchloroethylene, see Tetrachloroethylene
Phenformin (DBI, Meltrol), 95
Phenobarbital (Luminal), 89, 107, 108, 109, 110, 127
recovery of, 109
Phenol(s), 162, 163
detection limits of, 163
Phosphine, 66
Phosphoric acid, 163
Phosphorus pentoxide, 56, 117
Phosphotungstic acid, 114
Picric acid, 114
β-Pinene, 169
Piperitone, 169
Potassium carbonate, 28, 44, 57, 102, 175
Potassium dihydrogen phosphate, 170
Potassium ferricyanide, 22, 23, 35, 36
Potassium hydroxide, 120
Potassium pyrogallate, 26
5-Pregene-3β,16α-diol-20-one, 166
TMSi derivative, 166
5-Pregene-3β,21-diol-20-one, 166
TMSi derivative, 166
Pregnanediol, 165
Pregnanetriol, 165
Primidone (Mysoline), 128
Promazine (Sparine), 127
Propane, 19
Propane-1,2-diol, 157
n-Propanol, 103, 104, 105, 106, 112, 114, 115, 117, 119, 120, 123, 125, 131, 132, 136, 140, 142, 143, 145, 149, 177

Propionaldehyde, 119, 156, 169
Propionitrile, 119
n-Propylacetate, 102, 136, 178
Propylene dichloride, 66, 68
Propylene oxide, 64
n-Propylnitrite, 142, 143
Protein, 103, 110, 111, 114, 123, 140
  precipitation of, 103, 110, 111, 114, 123, 140
Protein binding, 38, 59, 61
  expressed in concentration terms, 59, 61
  expressed in partition coefficient terms, 61
  partition coefficient of, 61
    expressed as, 61
  percent bound, 59, 61
    equations for, 59, 61
Pulmonary capillary blood flow, 10, 11
Pulmonary capillary volume, 10, 11
Pulmonary clearance, 38
Pulmonary diffusing capacity, 10
Pulmonary "first pass," 38
Pyridine, 119
Pyrrole, 168, 169
Pyruvic acid, 157

Q

Quadrupole mass spectrometer, 133
Qualitative methods of analysis
  by elution temperature, 72, 73, 171, 174
  by integrated GC-MS, 75
  by mass spectra, 171
  by mass spectrometry, 48
  by relative retention time, 51, 66, 67, 69, 71, 72, 73, 117, 118, 119, 136, 171, 177, 178
[Qualitative methods of analysis]
  by retention time, 2, 4, 5, 6, 7, 14, 15, 17, 21, 26, 34, 36, 39, 40, 41, 42, 44, 48, 49, 50, 51, 56, 57, 60, 65, 66, 67, 68, 71, 72, 73, 74, 75, 88, 102, 104, 106, 115, 118, 120, 121, 123, 124, 129, 132, 136, 137, 140, 141, 142, 143, 144, 145, 149, 152, 153, 155, 157, 162, 163, 165, 166, 167, 170, 175, 178, 179
  by use of methylene units, 164
Quantitative methods of analysis
  by automatic integration, 11, 42, 55, 117, 121, 125, 126, 148
  by computer, 126, 128, 129, 133, 170, 171
  by integrated GC-MS, 128, 153, 155, 156, 157, 163, 166, 168, 171
  by integrated GC-MS-COMP, 128-129, 133-134, 171
    detection limits of, 134
  by multiple ion detection, 128-129, 133, 134
  by peak area measurement, 11, 12, 13, 24, 26, 27, 42, 55, 103, 105, 106, 112, 114, 117, 122, 124, 134, 148, 149, 150
  by peak height measurement, 3, 27, 31, 35, 36, 42, 55, 111, 115, 120, 124, 140, 148, 149, 150
  by specific ion detection (m/e), 128, 134
  by use of calibration curve, 27, 39, 51, 102, 111, 117, 120, 121, 122, 134, 143
  by use of internal standard, 64, 69, 71, 72, 103, 106, 107, 109, 110, 115, 117, 118, 120, 122, 123, 124, 125, 130, 131, 133, 134, 137, 140, 142, 144, 146, 148, 149, 152, 172, 177

Quinacrine, 94
Quinic acid, 164

R

Radium-226 beta ionization detector, see Detector, radium-226 beta ionization
Reaction gas chromatography, 115, 117
  definition of, 115, 117
  description of, 117
Red cell hemoglobin, 8
Relative response factor, 72
Respiratory gases, 3-21
  dual-column/dual-detector GC system for, 11, 12, 15, 19
    description of, 11, 12, 15
    schematic of, 15, 19
  dual-column/single-detector GC system for, 18, 19, 20

S

Salsolinol, 92, 93
  formation of, 92, 93
  structure, 93
Saponin, 22, 23, 35, 36
Secobarbital (Seconal), 108, 109, 110, 127
  recovery of, 109
Semicarbazide, 130
Serotonin, see 5-Hydroxytryptamine
Shikimic acid, 164
Silica gel, see Column packing, types of
Silver metal powder, 3
Smooth endoplasmic reticulum (hepatic), 89
  effect of ethyl alcohol on, 89
Sodium azide, 133
Sodium bicarbonate, 23
Sodium bisulfite, 102, 132, 133
Sodium carbonate, 28
Sodium chloride, 138, 140
Sodium dithionite, 133
Sodium fluoride, 133, 140, 143
Sodium hypochlorite, 114
Sodium nitrite, 133, 141
Sodium pyrophosphate, 130
Sodium sulfate, 111, 112
Sodium tungstate, 110, 111, 123
Solubility coefficient, 9
Specific gravity, 53
Specific ion detection (m/e), 128, 134
Stearic acid, 137
  N, N-dimethylamide, 137
Succinic acid, 167, 172
Succinyl choline, 45
Sulfur dioxide, 4
Sulfuric acid, 110, 111, 123, 152, 175
Sulphatase, 171
Support-coated open tubular column (SCOT), 131

T

Tartaric acid, 173
Temperature programming, 6, 18, 19, 20, 21, 64, 66, 68, 71, 74, 158, 164, 165, 166, 167, 168, 170, 174, 178, 179
Temposil, see Calcium carbimide
Terpene, 87
sym-Tetrachloroethane, 66, 67, 68
Tetrachloroethylene, 38, 54, 55, 56, 57, 66, 67, 70, 71, 136
  detection limits of, 55
Tetrachlorothiaphene, 69
  detection limits of, 69
Tetrahydrocortisol, 165
Tetrahydrocortisone, 165
Tetrahydro-11-dehydrocorticosterone, 165
Tetrahydropapaveroline (THP), 91, 92, 93

[Tetrahydropapaveroline (THP)]
formation of, 92, 93
structure, 93
Tetronic acid, 173
Thermal conductivity detector, see Detector, thermal conductivity
Thermistor detector, see Detector, thermal conductivity
Thiolan-2-one, 169
Thymol, 49, 131
Tissue analyses (volatiles), 171-180
Tolbutamide, 89, 94, 150
structure, 89
Toluene, 44, 69, 70, 71, 119, 124, 136, 169, 178
Transfer capacity of carbon monoxide, see Diffusing or diffusion capacity (lung), of carbon monoxide, defined as
Tricarboxylic acid cycle, 170
1,1,1-Trichloroethane, 38, 54, 55, 66, 70, 71, 119
detection limits of, 55
Trichloroethanol, 109, 110, 124, 141
Trichloroethylene, 1, 3, 4, 38, 39, 42, 43, 54, 55, 59, 66, 67, 69, 70, 71, 136, 175
detection limits of, 55
Trichlorofluoromethane (Freon 11), 56, 59, 60, 62, 63, 74, 75
detection limits of, 60
partition coefficient of, in, 60
1,1,2-Trichloro-1,2,2-trifluoroethane (Freon 113), 51, 52, 56, 60
detection limits of, 60
partition coefficient of, in, 60
Tridecane, 174
Trifluoroethyl vinyl ether, see Fluoromar
2,3,5-Trimethylfuran, 169
2,3,5-Trimethylpyrazine, 169

U

Undecane, 174
Urea, 141
Urinary acids, 170, 172-173
excretion rate of, 172-173
Urinary metabolic profile, 161, 163, 164, 165, 166, 168, 170, 171
Urinary steroids, 164, 165, 166, 171
MO-TMSi derivatives, 165
TMSi derivatives, 166
Urinary sulfur volatiles, 170
Urine analysis (volatiles), 161-171
Urine sample holder (volatiles), 160
schematic of, 160

V

Vapor pressure, 45, 138, 144, 150, 175
of common anesthetics, 45
Vigreaux column, 112
Vinylpyrazine, 169

W

Washout curve, 4, 8
definition of, 4
Water, 37, 86, 87, 115, 117, 118, 119, 120, 122, 123, 124, 131, 137, 177
ionization potential of, 37

X

Xenon, 21
Xylene, 70, 71, 136, 178
Xylylbromide (isomers), 72, 73

Z

Zinc sulfate, 140